100 Instructive Calculus-based

PHYSICS

Examples

Volume 1: The Laws of Motion

$$PE_0 + KE_0 + W_{nc} = PE + KE$$
$$mgh_0 = mgh + \frac{1}{2}mv^2$$
$$gh_0 = g2R + \frac{v^2}{2}$$

$$\sum F_{in} = ma_c$$
$$N + mg = ma_c$$
$$N = m(a_c - g)$$

$$gh_0 > g2R + \frac{Rg}{2}$$
$$h_0 > 2R + \frac{R}{2}$$
$$h_0 > \frac{4R}{2} + \frac{R}{2}$$
$$h_0 > \frac{5R}{2}$$

$$N > 0 \implies a_c > g$$
$$a_c > g \implies \frac{v^2}{R} > g$$
$$\frac{v^2}{R} > g \implies v^2 > Rg$$

 Chris McMullen, Ph.D.

100 Instructive Calculus-based Physics Examples
Volume 1: The Laws of Motion
Fundamental Physics Problems

Chris McMullen, Ph.D.
Physics Instructor
Northwestern State University of Louisiana

Copyright © 2016 Chris McMullen, Ph.D.
Updated edition: February, 2017

www.monkeyphysicsblog.wordpress.com
www.improveyourmathfluency.com
www.chrismcmullen.wordpress.com

Zishka Publishing

ISBN: 978-1-941691-17-5

Textbooks > Science > Physics
Study Guides > Workbooks> Science

CONTENTS

INTRODUCTION

This book includes fully-solved examples with detailed explanations for 139 standard physics problems. There are also 66 math examples, including algebra and calculus, which are essential toward mastering physics. That makes a total of 205 problems.

Each example breaks the solution down into terms that make it easy to understand. The written explanations between the math help describe exactly what is happening, one step at a time. These examples are intended to serve as a helpful guide for solving similar standard physics problems from a textbook or course.

The best way to use this book is to write down the steps of the mathematical solution on a separate sheet of paper while reading through the example. Since writing is a valuable memory aid, this is an important step. In addition to writing down the solution, try to think your way through the solution. It may help to read through the solution at least two times: The first time, write it down and work it out on a separate sheet of paper as you solve it. The next time, think your way through each step as you read it.

Math and science books aren't meant to be read like novels. The best way to learn math and science is to think it through one step at a time. Read an idea, think about it, and then move on. Also write down the solutions and work them out on your own paper as you read. Students who do this tend to learn math and science better.

Note that these examples serve two purposes:

- They are primarily designed to help students understand how to solve standard physics problems. This can aid students who are struggling to figure out homework problems, or it can help students prepare for exams.
- These examples are also the solutions to the problems of the author's other book, *Essential Calculus-based Physics Study Guide Workbook*, ISBN 978-1-941691-15-1. That study guide workbook includes space on which to solve each problem.

1 REVIEW OF ESSENTIAL ALGEBRA SKILLS

We begin with a review of essential algebra skills. If you feel confident working with fractional powers, factoring, applying the quadratic equation, and solving systems of equations, you may move onto the next chapter.

Note: The math examples in this book are **not** counted as part of the 100 physics examples. There are over 100 problems in this book devoted solely to physics.

Example 1. Simplify the following expression.
$$x^2 x^3 x^4$$
Solution. Recall the rule from algebra that $x^m x^n = x^{m+n}$. For this problem, we extend this rule to $x^l x^m x^n = x^{l+m+n}$. This means that we can simply add the exponents together.
$$x^2 x^3 x^4 = x^{2+3+4} = x^9$$
The answer is x^9.

Example 2. Simplify the following expression.
$$\frac{x^4 x^5}{x^6}$$
Solution. First apply the rule $x^m x^n = x^{m+n}$ to simplify the numerator. When multiplying powers of the same base, add the exponents together.
$$x^4 x^5 = x^{4+5} = x^9$$
Now apply the rule $\frac{x^m}{x^n} = x^{m-n}$. When dividing powers of the same base, subtract the exponents.
$$\frac{x^9}{x^6} = x^{9-6} = x^3$$
From start to finish, the complete solution is:
$$\frac{x^4 x^5}{x^6} = \frac{x^{4+5}}{x^6} = \frac{x^9}{x^6} = x^{9-6} = x^3$$
The answer is x^3.

Example 3. Simplify the following expression.
$$\frac{x^{-7}}{x^{-8}}$$
Solution. Apply the rule $\frac{x^m}{x^n} = x^{m-n}$. Subtract the exponents. When you subtract a negative number, the two minus signs make a plus sign.
$$\frac{x^{-7}}{x^{-8}} = x^{-7-(-8)} = x^{-7+8} = x^1 = x$$
Recall from algebra that $x^1 = x$. The answer is x.

Example 4. Simplify the following expression.

$$\frac{x^3 x^{-6}}{x^{-2} x^5}$$

Solution. First apply the rule $x^m x^n = x^{m+n}$ to simplify the numerator. When multiplying powers of the same base, add the exponents together.

$$x^3 x^{-6} = x^{3-6} = x^{-3}$$

Next apply the same rule to simplify the denominator.

$$x^{-2} x^5 = x^{-2+5} = x^3$$

Now apply the rule $\frac{x^m}{x^n} = x^{m-n}$ to combine the previous results together. When dividing powers of the same base, subtract the exponents.

$$\frac{x^3 x^{-6}}{x^{-2} x^5} = \frac{x^{-3}}{x^3} = x^{-3-3} = x^{-6}$$

The answer is x^{-6}. Note that the answer is the same as $\frac{1}{x^6}$.

Example 5. Simplify the following expression.

$$x\sqrt{x}$$

Solution. First rewrite x and $\sqrt{x}$ as powers of x. Recall that $x^1 = x$ (a power of 1 is implied when you don't see an exponent). Also recall the rule that $x^{1/2} = \sqrt{x}$.

$$x\sqrt{x} = x^1 x^{1/2}$$

Now apply the rule $x^m x^n = x^{m+n}$. When multiplying powers of the same base, add the exponents together.

$$x^1 x^{1/2} = x^{1+1/2}$$

Find a **common denominator** in order to add the fractions in the exponent.

$$1 + \frac{1}{2} = \frac{2}{2} + \frac{1}{2} = \frac{2+1}{2} = \frac{3}{2}$$
$$x^{1+1/2} = x^{3/2}$$

From start to finish, the complete solution is:

$$x\sqrt{x} = x^1 x^{1/2} = x^{1+1/2} = x^{3/2}$$

The answer is $x^{3/2}$.

Example 6. Simplify the following expression.

$$\frac{x^{3/2}}{\sqrt{x}}$$

Solution. Apply the rules $x^{1/2} = \sqrt{x}$ and $\frac{x^m}{x^n} = x^{m-n}$. Subtract the exponents.

$$\frac{x^{3/2}}{\sqrt{x}} = \frac{x^{3/2}}{x^{1/2}} = x^{3/2-1/2} = x^1 = x$$

Note that $\frac{3}{2} - \frac{1}{2} = 1$. Recall that $x^1 = x$. The answer is x.

Example 7. Write the following expression in standard form.
$$(4x)^{-1/2}$$

Solution. First apply the rule $(ax)^{-1/2} = \frac{1}{\sqrt{ax}}$. Note that this rule follows from the rules $(ax)^m = a^m x^m$ and $x^{-1/2} = \frac{1}{x^{1/2}} = \frac{1}{\sqrt{x}}$.
$$(4x)^{-1/2} = \frac{1}{\sqrt{4x}}$$

This expression isn't in standard form because the denominator has a squareroot. In order to put it in standard form, we must **rationalize the denominator.** Multiply the numerator and denominator by $\sqrt{4x}$. Note that $\sqrt{4x}\sqrt{4x} = \sqrt{4^2 x^2} = 4x$ (any squareroot multiplied by itself effectively removes the squareroot, as in $\sqrt{x}\sqrt{x} = x$).
$$\frac{1}{\sqrt{4x}} = \frac{1}{\sqrt{4x}}\frac{\sqrt{4x}}{\sqrt{4x}} = \frac{\sqrt{4x}}{4x}$$

Simplify the numerator using the rule $\sqrt{ax} = \sqrt{a}\sqrt{x}$.
$$\sqrt{4x} = \sqrt{4}\sqrt{x} = 2\sqrt{x}$$

Plug this in to the previous expression and simplify.
$$\frac{\sqrt{4x}}{4x} = \frac{2\sqrt{x}}{4x} = \frac{2}{4}\frac{\sqrt{x}}{x} = \frac{\sqrt{x}}{2x}$$

The complete solution is:
$$(4x)^{-1/2} = \frac{1}{\sqrt{4x}} = \frac{1}{\sqrt{4x}}\frac{\sqrt{4x}}{\sqrt{4x}} = \frac{\sqrt{4x}}{4x} = \frac{2\sqrt{x}}{4x} = \frac{\sqrt{x}}{2x}$$

The answer is $\frac{\sqrt{x}}{2x}$.

Example 8. Subtract the following fractions.
$$\frac{2}{x^2} - \frac{3}{x^3}$$

Solution. In order to subtract fractions, make a **common denominator**. The denominator x^2 can be turned into x^3 by multiplying by x. That is, $x^2 x = x^2 x^1 = x^{2+1} = x^3$ (since $x^1 = x$). Multiply $\frac{2}{x^2}$ by $\frac{x}{x}$ in order to make the common denominator of x^3.
$$\frac{2}{x^2} = \frac{2}{x^2}\frac{x}{x} = \frac{2x}{x^3}$$

Once we have a common denominator, then we may subtract the numerators. Following is the complete solution:
$$\frac{2}{x^2} - \frac{3}{x^3} = \frac{2}{x^2}\frac{x}{x} - \frac{3}{x^3} = \frac{2x}{x^3} - \frac{3}{x^3} = \frac{2x - 3}{x^3}$$

The answer is $\frac{2x-3}{x^3}$.

Example 9. Add the following expressions together.

$$\frac{1}{\sqrt{x}} + \sqrt{x}$$

Solution. Recall that anything divided by 1 equals itself. For example, $\frac{x}{1} = x$. The 'trick' to this problem is to rewrite $\sqrt{x}$ as $\frac{\sqrt{x}}{1}$, which we can do since $\frac{\sqrt{x}}{1} = \sqrt{x}$.

$$\frac{1}{\sqrt{x}} + \sqrt{x} = \frac{1}{\sqrt{x}} + \frac{\sqrt{x}}{1}$$

Now the problem looks like the addition of two fractions. In order to add fractions, make a **common denominator**. The denominator 1 can be turned into $\sqrt{x}$ by multiplying by $\sqrt{x}$. That is, $1\sqrt{x} = \sqrt{x}$ (since anything multiplied by 1 equals itself). Multiply $\frac{\sqrt{x}}{1}$ by $\frac{\sqrt{x}}{\sqrt{x}}$.

$$\frac{\sqrt{x}}{1} = \frac{\sqrt{x}}{1}\frac{\sqrt{x}}{\sqrt{x}} = \frac{x}{\sqrt{x}}$$

In the last step, we applied the rule $\sqrt{x}\sqrt{x} = x$. Once we have a common denominator, then we may add the numerators. Following, we combine the previous steps together.

$$\frac{1}{\sqrt{x}} + \sqrt{x} = \frac{1}{\sqrt{x}} + \frac{\sqrt{x}}{1} = \frac{1}{\sqrt{x}} + \frac{\sqrt{x}}{1}\frac{\sqrt{x}}{\sqrt{x}} = \frac{1}{\sqrt{x}} + \frac{x}{\sqrt{x}} = \frac{1+x}{\sqrt{x}} = \frac{x+1}{\sqrt{x}}$$

Note that $1 + x$ is the same thing as $x + 1$, since addition is commutative. We changed the order because it's conventional to write the variable before the constant. Our answer is not yet in standard form. In order to express the answer in standard form, we must first **rationalize the denominator.** Multiply the numerator and denominator by $\sqrt{x}$.

$$\frac{x+1}{\sqrt{x}} = \frac{x+1}{\sqrt{x}}\frac{\sqrt{x}}{\sqrt{x}} = \frac{(x+1)\sqrt{x}}{x}$$

Distribute the $\sqrt{x}$ in the numerator.

$$(x+1)\sqrt{x} = x\sqrt{x} + \sqrt{x}$$

Applying this to the previous equation, we get

$$\frac{x+1}{\sqrt{x}} = \frac{x\sqrt{x} + \sqrt{x}}{x}$$

The answer is $\frac{x\sqrt{x}+\sqrt{x}}{x}$.

Example 10. Apply the distributive property to the following expression.

$$6x(x + 9)$$

Solution. Distribute the $6x$ to the two terms, which are x and 9. Note that $xx = x^1x^1 = x^2$.

$$6x(x + 9) = 6x(x) + 6x(9) = 6x^2 + 54x$$

The answer is $6x^2 + 54x$.

Example 11. Apply the distributive property to the following expression.
$$-3x^2(5x^6 - 2x^4)$$
Solution. Distribute the $-3x^2$ to the two terms, which are $5x^6$ and $-2x^4$.
$$-3x^2(5x^6 - 2x^4) = -3x^2(5x^6) - 3x^2(-2x^4)$$
Two minus signs make a plus sign.
$$-3x^2(-2x^4) = 3x^2(2x^4)$$
Apply this property of minus signs to the original equation.
$$-3x^2(5x^6) - 3x^2(-2x^4) = -3x^2(5x^6) + 3x^2(2x^4)$$
Apply the rule $x^m x^n = x^{m+n}$.
$$-3x^2(5x^6) + 3x^2(2x^4) = -15x^{2+6} + 6x^{2+4} = -15x^8 + 6x^6$$
The answer is $-15x^8 + 6x^6$.

Example 12. Apply the distributive property to the following expression.
$$\sqrt{x}(x + \sqrt{x})$$
Solution. Distribute the $\sqrt{x}$ to the two terms, which are x and $\sqrt{x}$.
$$\sqrt{x}(x + \sqrt{x}) = \sqrt{x}(x) + \sqrt{x}\sqrt{x}$$
Apply the rule $\sqrt{x}\sqrt{x} = x$.
$$\sqrt{x}(x) + \sqrt{x}\sqrt{x} = \sqrt{x}(x) + x$$
Apply the rules $x^{1/2} = \sqrt{x}$ and $x^1 = x$.
$$\sqrt{x}(x) + x = x^{1/2}x^1 + x$$
Apply the rule $x^m x^n = x^{m+n}$. When multiplying powers of the same base, add the exponents together.
$$x^1 x^{1/2} + x = x^{1+1/2} + x$$
Find a **common denominator** in order to add the fractions in the exponent.
$$1 + \frac{1}{2} = \frac{2}{2} + \frac{1}{2} = \frac{2+1}{2} = \frac{3}{2}$$
Apply this arithmetic to the previous equation.
$$x^{1/2}x^1 + x = x^{3/2} + x$$
The answer is $x^{3/2} + x$.

Example 13. Apply the foil method to the following expression.
$$(3x + 2)(x + 5)$$
Solution. Recall the foil method. The word "foil" is an acronym (first, outside, inside, last): Multiply the first terms together $3x(x)$, multiply the outside terms together $3x(5)$, multiply the inside terms together $2(x)$, and multiply the last terms together $2(5)$.
$$(3x + 2)(x + 5) = 3x(x) + 3x(5) + 2(x) + 2(5) = 3x^2 + 15x + 2x + 10$$
$$(3x + 2)(x + 5) = 3x^2 + 17x + 10$$
In the last step, we combined like terms together: The $15x$ and $2x$ are like terms. They combine to make $15x + 2x = 17x$. The answer is $3x^2 + 17x + 10$.

Example 14. Apply the foil method to the following expression.
$$(6x - 4)^2$$
Solution. First rewrite $(6x - 4)^2$ as $(6x - 4)$ times itself.
$$(6x - 4)^2 = (6x - 4)(6x - 4)$$
As explained in the previous example, the foil method involves four steps: Multiply the first terms together $6x(6x)$, multiply the outside terms together $6x(-4)$, multiply the inside terms together $-4(6x)$, and multiply the last terms together $-4(-4)$.
$$(6x - 4)(6x - 4) = 6x(6x) + 6x(-4) - 4(6x) - 4(-4) = 36x^2 - 24x - 24x + 16$$
$$(6x - 4)(6x - 4) = 36x^2 - 48x + 16$$
In the last step, we combined like terms together: The $-24x$ and $-24x$ are like terms. They combine to make $-24x - 24x = -48x$. The answer is $36x^2 - 48x + 16$.

Example 15. Factor the following expression:
$$x^3 - 4x$$
Solution. Factor an x out of each expression.
- Write the first term as $x^3 = x(x^2)$.
- Write the second term as $-4x = x(-4)$.

Factor out the x:
$$x^3 - 4x = x(x^2) + x(-4) = x(x^2 - 4)$$
The answer is $x(x^2 - 4)$.
Check. We can check our answer by distributing.
$$x(x^2 - 4) = x(x^2) + x(-4) = x^1 x^2 - 4x = x^{1+2} - 4x = x^3 - 4x \checkmark$$
We applied the rule $x^m x^n = x^{m+n}$. When multiplying powers of the same base, add the exponents together.

Example 16. Factor the following expression:
$$8x^9 + 12x^6$$
Solution. Factor $4x^6$ out of each expression. Apply the rule $x^m x^n = x^{m+n}$. When multiplying powers of the same base, add the exponents together.
- Write the first term as $8x^9 = (4)(2)x^6 x^3 = 4x^6(2x^3)$.
- Write the second term as $12x^6 = (4)(3)x^6 = 4x^6(3)$.

Factor out the $4x^6$:
$$8x^9 + 12x^6 = 4x^6(2x^3) + 4x^6(3) = 4x^6(2x^3 + 3)$$
The answer is $4x^6(2x^3 + 3)$.
Check. We can check our answer by distributing.
$$4x^6(2x^3 + 3) = 4x^6(2x^3) + 4x^6(3) = (4)(2)x^{6+3} + (4)(3)x^6 = 8x^9 + 12x^6 \checkmark$$

Example 17. Factor the following expression:
$$45x^7 - 18x^5 + 27x^3$$
Solution. Factor $9x^3$ out of each expression. Apply the rule $x^m x^n = x^{m+n}$. When multiplying powers of the same base, add the exponents together.

- Write the first term as $45x^7 = (9)(5)x^3 x^4 = 9x^3(5x^4)$.
- Write the second term as $-18x^5 = (9)(-2)x^3 x^2 = 9x^3(-2x^2)$.
- Write the last term as $27x^3 = (9)(3)x^3 = 9x^3(3)$

Factor out the $9x^3$:
$$45x^7 - 18x^5 + 27x^3 = 9x^3(5x^4) + 9x^3(-2x^2) + 9x^3(3) = 9x^3(5x^4 - 2x^2 + 3)$$
The answer is $9x^3(5x^4 - 2x^2 + 3)$.

Check. We can check our answer by distributing.
$$9x^3(5x^4 - 2x^2 + 3) = 9x^3(5x^4) + 9x^3(-2x^2) + 9x^3(3)$$
$$= (9)(5)x^{3+4} - (9)(2)x^{3+2} + 9(3)x^3 = 45x^7 - 18x^5 + 27x^3 \checkmark$$

Example 18. Express the following irrational number in standard form.
$$\sqrt{18}$$
Solution. The number $\sqrt{18}$ isn't in standard form because we can factor out a perfect square. List the **perfect squares** up to 18.
$$2^2 = 4 \quad , \quad 3^2 = 9 \quad , \quad 4^2 = 16$$
What is the largest perfect square that factors into 18? It's 9 because we can write 18 as $(9)(2)$. Apply the rule $\sqrt{ax} = \sqrt{a}\sqrt{x}$.
$$\sqrt{18} = \sqrt{(9)(2)} = \sqrt{9}\sqrt{2} = 3\sqrt{2}$$
Note that $\sqrt{9} = 3$. The answer is $3\sqrt{2}$.

Example 19. Express the following irrational number in standard form.
$$\sqrt{108}$$
Solution. The number $\sqrt{108}$ isn't in standard form because we can factor out a perfect square. List the **perfect squares** up to 108.
$$2^2 = 4 \quad , \quad 3^2 = 9 \quad , \quad 4^2 = 16 \quad , \quad 5^2 = 25 \quad , \quad 6^2 = 36$$
$$7^2 = 49 \quad , \quad 8^2 = 64 \quad , \quad 9^2 = 81 \quad , \quad 10^2 = 100$$
What is the largest perfect square that factors into 108? It's 36 because we can write 108 as $(36)(3)$. Apply the rule $\sqrt{ax} = \sqrt{a}\sqrt{x}$.
$$\sqrt{108} = \sqrt{(36)(3)} = \sqrt{36}\sqrt{3} = 6\sqrt{3}$$
Note that $\sqrt{36} = 6$. The answer is $6\sqrt{3}$.

Example 20. Use the quadratic formula to solve for x in the following equation.
$$2x^2 - 2x - 40 = 0$$
Solution. This is a quadratic equation because it has one term with the variable **squared** ($2x^2$ contains x^2), one **linear** term ($-2x$ is linear because it includes x), and one **constant** term (-40 is constant because it doesn't include a variable). The **standard form** of the quadratic equation is:
$$ax^2 + bx + c = 0$$
Note that the given equation is already in standard form, since the squared term ($2x^2$) is first, the linear term ($-2x$) is second, and the constant term (-40) is last. Identify the constants a, b, and c by comparing $2x^2 - 2x - 40 = 0$ with $ax^2 + bx + c = 0$.
$$a = 2 \quad , \quad b = -2 \quad , \quad c = -40$$
Note that b and c are both negative. Plug these values into the quadratic formula.
$$x = \frac{-b \pm \sqrt{b^2 - 4ac}}{2a} = \frac{-(-2) \pm \sqrt{(-2)^2 - 4(2)(-40)}}{2(2)}$$
Note that $-(-2) = +2$ since the two minus signs make a plus sign. Also note that the minus sign doesn't matter in $(-2)^2$, since the minus sign gets squared: $(-2)^2 = +4$. It's a common mistake for students to incorrectly type -2^2 or $-(2)^2$ on their calculator when the correct thing to type is $(-2)^2$, which is the same thing as 2^2. Similarly, the two minus signs inside the squareroot make a plus sign: $-4(2)(-40) = +320$.
$$x = \frac{2 \pm \sqrt{4 + 320}}{4} = \frac{2 \pm \sqrt{324}}{4} = \frac{2 \pm 18}{4}$$
There are two solutions for x. We must work these out separately.
$$x = \frac{2 + 18}{4} \quad \text{or} \quad x = \frac{2 - 18}{4}$$
$$x = \frac{20}{4} \quad \text{or} \quad x = \frac{-16}{4}$$
$$x = 5 \quad \text{or} \quad x = -4$$
The two answers are $x = 5$ and $x = -4$.
Check. We can check our answers by plugging them into the original equation.
$$2x^2 - 2x - 40 = 2(5)^2 - 2(5) - 40 = 2(25) - 10 - 40 = 50 - 50 = 0 \checkmark$$
$$2x^2 - 2x - 40 = 2(-4)^2 - 2(-4) - 40 = 2(16) + 8 - 40 = 32 + 8 - 40 = 0 \checkmark$$

Example 21. Use the quadratic formula to solve for y.
$$3y - 27 + 2y^2 = 0$$
Solution. Reorder the terms in **standard form**. Put the squared term ($2y^2$) first, the linear term ($3y$) next, and the constant term (-27) last.
$$2y^2 + 3y - 27 = 0$$
Identify the constants a, b, and c by comparing $2y^2 + 3y - 27 = 0$ with $ay^2 + by + c = 0$.
$$a = 2 \quad , \quad b = 3 \quad , \quad c = -27$$
Plug these values into the quadratic formula.

$$y = \frac{-b \pm \sqrt{b^2 - 4ac}}{2a} = \frac{-3 \pm \sqrt{3^2 - 4(2)(-27)}}{2(2)}$$

Note that the two minus signs make a plus sign: $-4(2)(-27) = +216$.

$$y = \frac{-3 \pm \sqrt{9 + 216}}{4} = \frac{-3 \pm \sqrt{225}}{4} = \frac{-3 \pm 15}{4}$$

$$y = \frac{-3 + 15}{4} \quad \text{or} \quad y = \frac{-3 - 15}{4}$$

$$y = \frac{12}{4} \quad \text{or} \quad y = \frac{-18}{4}$$

Note that $-\frac{18}{4}$ reduces to $-\frac{9}{2}$ if you divide both the numerator and denominator by 2. That is, $-\frac{18}{4} = -\frac{18 \div 2}{4 \div 2} = -\frac{9}{2}$. Simplifying the previous equations, we get:

$$y = 3 \quad \text{or} \quad y = -\frac{9}{2}$$

The two answers are $y = 3$ and $y = -\frac{9}{2}$.

Check. We can check our answers by plugging them into the original equation.

$$3y - 27 + 2y^2 = 3(3) - 27 + 2(3)^2 = 9 - 27 + 2(9) = 9 - 27 + 18 = 0 \checkmark$$

$$3y - 27 + 2y^2 = 3\left(-\frac{9}{2}\right) - 27 + 2\left(\frac{9}{2}\right)^2 = -\frac{27}{2} - 27 + 2\left(\frac{81}{4}\right)$$

To add or subtract fractions, make a common denominator. We can make a common denominator of 4 by multiplying $-\frac{27}{2}$ by $\frac{2}{2}$ and multiplying -27 by $\frac{4}{4}$.

$$-\frac{27}{2} - 27 + 2\left(\frac{81}{4}\right) = -\frac{27}{2}\left(\frac{2}{2}\right) - 27\left(\frac{4}{4}\right) + 2\left(\frac{81}{4}\right)$$

$$= -\frac{54}{4} - \frac{108}{4} + \frac{162}{4} = \frac{-54 - 108 + 162}{4} = 0 \checkmark$$

Example 22. Use the quadratic formula to solve for t.

$$6t = 8 - 2t^2$$

Solution. Reorder the terms in **standard form**. Use algebra to bring all of the terms to the same side of the equation (we will put them on the left side). Put the squared term $(-2t^2)$ first, the linear term $(6t)$ next, and the constant term (8) last. Note that the sign of a term will change if it is brought from the right-hand side to the left-hand side (we're subtracting 8 from both sides and we're adding $2t^2$ to both sides of the equation).

$$2t^2 + 6t - 8 = 0$$

Identify the constants a, b, and c by comparing $2t^2 + 6t - 8 = 0$ with $at^2 + bt + c = 0$.

$$a = 2 \quad , \quad b = 6 \quad , \quad c = -8$$

Plug these values into the quadratic formula.

$$t = \frac{-b \pm \sqrt{b^2 - 4ac}}{2a} = \frac{-6 \pm \sqrt{6^2 - 4(2)(-8)}}{2(2)}$$

Note that the two minus signs make a plus sign: $-4(2)(-8) = +64$.

$$t = \frac{-6 \pm \sqrt{36 + 64}}{4} = \frac{-6 \pm \sqrt{100}}{4} = \frac{-6 \pm 10}{4}$$

$$t = \frac{-6 + 10}{4} \quad \text{or} \quad t = \frac{-6 - 10}{4}$$

$$t = \frac{4}{4} \quad \text{or} \quad t = \frac{-16}{4}$$

$$t = 1 \quad \text{or} \quad t = -4$$

The two answers are $t = 1$ and $t = -4$.

Check. We can check our answers by plugging them into the original equation. For each answer, we'll compare the left-hand side ($6t$) with the right-hand side ($8 - 2t^2$). First plug $t = 1$ into both sides of $6t = 8 - 2t^2$.

$$6t = 6(1) = 6 \quad \text{and} \quad 8 - 2t^2 = 8 - 2(1)^2 = 8 - 2 = 6 \checkmark$$

Now plug in $t = -4$ into both sides of $6t = 8 - 2t^2$.

$$6t = 6(-4) = -24 \quad \text{and} \quad 8 - 2t^2 = 8 - 2(-4)^2 = 8 - 2(16) = 8 - 32 = -24 \checkmark$$

Example 23. Use the quadratic formula to solve for x.

$$1 + 25x - 5x^2 = 8x - 3x^2 + 9$$

Solution. Reorder the terms in **standard form**. Use algebra to bring all of the terms to the same side of the equation (we will put them on the left side). Put the squared terms ($-5x^2$ and $-3x^2$) first, the linear terms ($25x$ and $8x$) next, and the constant terms (1 and 9) last. Note that the sign of a term will change if it is brought from the right-hand side to the left-hand side (we're subtracting $8x$ from both sides, adding $3x^2$ to both sides, and subtracting 9 from both sides of the equation).

$$-5x^2 + 3x^2 + 25x - 8x + 1 - 9 = 0$$

Combine like terms: Combine the two x^2 terms, the two x terms, and the two constants.

$$-2x^2 + 17x - 8 = 0$$

Identify the constants a, b, and c by comparing $-2x^2 + 17x - 8 = 0$ with $ax^2 + bx + c = 0$.

$$a = -2 \quad , \quad b = 17 \quad , \quad c = -8$$

Plug these values into the quadratic formula.

$$x = \frac{-b \pm \sqrt{b^2 - 4ac}}{2a} = \frac{-17 \pm \sqrt{17^2 - 4(-2)(-8)}}{2(-2)}$$

Note that the three minus signs make a minus sign: $-4(-2)(-8) = -64$.

$$x = \frac{-17 \pm \sqrt{289 - 64}}{-4} = \frac{-17 \pm \sqrt{225}}{-4} = \frac{-17 \pm 15}{-4}$$

$$x = \frac{-17 + 15}{-4} \quad \text{or} \quad x = \frac{-17 - 15}{-4}$$

$$x = \frac{-2}{-4} \quad \text{or} \quad x = \frac{-32}{-4}$$

Note that a negative number divided by a negative number results in a positive answer. The minus signs from the numerator and denominator cancel out.

$$x = \frac{1}{2} \quad \text{or} \quad x = 8$$

The two answers are $x = \frac{1}{2}$ and $x = 8$.

Check. We can check our answers by plugging them into the original equation. For each answer, we'll compare the left-hand side $(1 + 25x - 5x^2)$ with the right-hand side $(8x - 3x^2 + 9)$. First plug $x = \frac{1}{2}$ into the left-hand side $(1 + 25x - 5x^2)$.

$$1 + 25x - 5x^2 = 1 + 25\left(\frac{1}{2}\right) - 5\left(\frac{1}{2}\right)^2 = 1 + \frac{25}{2} - \frac{5}{4}$$

In order to add and subtract the fractions, multiply 1 by $\frac{4}{4}$ and multiply $\frac{25}{2}$ by $\frac{2}{2}$ to make a common denominator.

$$1 + \frac{25}{2} - \frac{5}{4} = 1\left(\frac{4}{4}\right) + \frac{25}{2}\frac{2}{2} - \frac{5}{4} = \frac{4}{4} + \frac{50}{4} - \frac{5}{4} = \frac{4 + 50 - 5}{4} = \frac{49}{4}$$

Next plug $x = \frac{1}{2}$ into the right-hand side $(8x - 3x^2 + 9)$.

$$8x - 3x^2 + 9 = 8\left(\frac{1}{2}\right) - 3\left(\frac{1}{2}\right)^2 + 9 = 4 - 3\left(\frac{1}{4}\right) + 9 = 13 - \frac{3}{4}$$

In order to subtract the fraction, multiply 13 by $\frac{4}{4}$ to make a common denominator.

$$13 - \frac{3}{4} = 13\left(\frac{4}{4}\right) - \frac{3}{4} = \frac{52}{4} - \frac{3}{4} = \frac{52 - 3}{4} = \frac{49}{4} \checkmark$$

Now plug in $x = 8$ into both sides of $1 + 25x - 5x^2 = 8x - 3x^2 + 9$.

$$1 + 25x - 5x^2 = 1 + 25(8) - 5(8)^2 = 1 + 200 - 5(64) = 201 - 320 = -119$$
$$8x - 3x^2 + 9 = 8(8) - 3(8)^2 + 9 = 64 - 3(64) + 9 = 64 - 192 + 9 = -119 \checkmark$$

Example 24. Use the method of substitution to solve the following system of equations for each unknown.

$$3x + 2y = 18$$
$$8x - 5y = 17$$

Solution. First isolate y in the top equation. Subtract $3x$ from both sides.

$$2y = 18 - 3x$$

Divide both sides of the equation by 2.

$$y = \frac{18 - 3x}{2}$$

Substitute this expression in parentheses in place of y in the bottom equation.

$$8x - 5y = 17$$
$$8x - 5\left(\frac{18 - 3x}{2}\right) = 17$$

Distribute the 5. When you distribute, the two minus signs make a plus.

$$8x - 5\left(\frac{18}{2}\right) - 5\left(-\frac{3x}{2}\right) = 17$$

$$8x - 5(9) + 5\left(\frac{3x}{2}\right) = 17$$

$$8x - 45 + \frac{15x}{2} = 17$$

Combine like terms: $8x$ and $\frac{15x}{2}$ are like terms, and -45 and 17 are like terms. Combine the terms $8x + \frac{15x}{2}$ using a **common denominator**. Multiply $8x$ by $\frac{2}{2}$. In order to combine the constant terms, add 45 to both sides.

$$8x\left(\frac{2}{2}\right) + \frac{15x}{2} = 17 + 45$$

$$\frac{16x}{2} + \frac{15x}{2} = 62$$

$$\frac{31x}{2} = 62$$

Multiply both sides of the equation by 2.

$$31x = 124$$

Divide both sides of the equation by 31.

$$x = 4$$

Now that we have an answer for x, we can plug it into one of the previous equations in order to solve for y. It's convenient to use the equation where y was isolated.

$$y = \frac{18 - 3x}{2} = \frac{18 - 3(4)}{2} = \frac{18 - 12}{2} = \frac{6}{2} = 3$$

The answers are $x = 4$ and $y = 3$.

Check. We can check our answers by plugging them into the original equations.

$$3x + 2y = 3(4) + 2(3) = 12 + 6 = 18 \checkmark$$

$$8x - 5y = 8(4) - 5(3) = 32 - 15 = 17 \checkmark$$

Example 25. Use the method of substitution to solve the following system of equations for each unknown.

$$4y + 3z = 10$$

$$5y - 2z = -22$$

Solution. First isolate y in the top equation. Subtract $3z$ from both sides.

$$4y = 10 - 3z$$

Divide both sides of the equation by 4.

$$y = \frac{10 - 3z}{4}$$

Substitute this expression in parentheses in place of y in the bottom equation.

$$5y - 2z = -22$$

$$5\left(\frac{10 - 3z}{4}\right) - 2z = -22$$

Distribute the 5.

$$5\left(\frac{10}{4}\right) - 5\left(\frac{3z}{4}\right) - 2z = -22$$

$$\frac{50}{4} - \frac{15z}{4} - 2z = -22$$

Combine like terms: $-\frac{15z}{4}$ and $-2z$ are like terms, and $\frac{50}{4}$ and -22 are like terms. Combine like terms using a **common denominator**. Multiply $-2z$ and -22 each by $\frac{4}{4}$ to make a common denominator. In order to combine the constant terms, subtract $\frac{50}{4}$ from both sides.

$$-\frac{15z}{4} - 2z\left(\frac{4}{4}\right) = -22\left(\frac{4}{4}\right) - \frac{50}{4}$$

$$-\frac{15z}{4} - \frac{8z}{4} = -\frac{88}{4} - \frac{50}{4}$$

$$\frac{-15z - 8z}{4} = \frac{-88 - 50}{4}$$

$$-\frac{23z}{4} = -\frac{138}{4}$$

Multiply both sides of the equation by 4.

$$-23z = -138$$

Divide both sides of the equation by -23. The two minus signs will cancel.

$$z = 6$$

Now that we have an answer for z, we can plug it into one of the previous equations in order to solve for y. It's convenient to use the equation where y was isolated.

$$y = \frac{10 - 3z}{4} = \frac{10 - 3(6)}{4} = \frac{10 - 18}{4} = \frac{-8}{4} = -2$$

The answers are $z = 6$ and $y = -2$.

Check. We can check our answers by plugging them into the original equations.

$$4y + 3z = 4(-2) + 3(6) = -8 + 18 = 10 \checkmark$$

$$5y - 2z = 5(-2) - 2(6) = -10 - 12 = -22 \checkmark$$

Example 26. Use the method of substitution to solve the following system of equations for each unknown.

$$3x - 4y + 2z = 44$$

$$5y + 6z = 29$$

$$2x + z = 13$$

Solution. It would be easiest to isolate z in the bottom equation. Subtract $2x$ from both sides in the bottom equation.

$$z = 13 - 2x$$

Substitute this expression in parentheses in place of z in the top two equations.

$$3x - 4y + 2(13 - 2x) = 44$$

$$5y + 6(13 - 2x) = 29$$

Distribute the 2 in the first equation and distribute the 6 in the second equation.

$$3x - 4y + 26 - 4x = 44$$
$$5y + 78 - 12x = 29$$

Combine like terms: These include $3x$ and $-4x$ in the first equation, 26 and 44 in the first equation, and 78 and 29 in the second equation. Reorder terms in the first equation to put like terms together. Subtract 26 from both sides in the first equation and subtract 78 from both sides in the second equation.

$$3x - 4x - 4y = 44 - 26$$
$$5y - 12x = 29 - 78$$

Simplify each equation. Note that $3x - 4x = -x$.

$$-x - 4y = 18$$
$$5y - 12x = -49$$

Now we have two equations in two unknowns (x and y), similar to the two previous examples. Isolate x in the first equation. Add $4y$ to both sides.

$$-x = 18 + 4y$$

Multiply both sides of the equation by -1.

$$x = -18 - 4y$$

Substitute this expression in parentheses in place of x in the final equation.

$$5y - 12x = -49$$
$$5y - 12(-18 - 4y) = -49$$

Distribute the 12. When you distribute, the two minus signs make a plus.

$$5y - 12(-18) - 12(-4y) = -49$$
$$5y + 12(18) + 12(4y) = -49$$
$$5y + 216 + 48y = -49$$

Combine like terms: $5y$ and $48y$ are like terms, and 216 and -49 are like terms. Subtract 216 from both sides.

$$5y + 48y = -49 - 216$$
$$53y = -265$$

Divide both sides by 53.

$$y = -5$$

Now that we have an answer for y, we can plug it into one of the previous equations in order to solve for x. It's convenient to use the equation where x was isolated.

$$x = -18 - 4y = -18 - 4(-5) = -18 + 4(5) = -18 + 20 = 2$$

Now that we have solves for x and y, we can plug them into one of the previous equations in order to solve for z. It's convenient to use the equation where z was isolated.

$$z = 13 - 2x = 13 - 2(2) = 13 - 4 = 9$$

The answers are $x = 2$, $y = -5$, and $z = 9$.

Check. We can check our answers by plugging them into the original equations.

$$3x - 4y + 2z = 3(2) - 4(-5) + 2(9) = 6 + 20 + 18 = 44 \ \checkmark$$
$$5y + 6z = 5(-5) + 6(9) = -25 + 54 = 29 \ \checkmark$$
$$2x + z = 2(2) + 9 = 4 + 9 = 13 \ \checkmark$$

2 REVIEW OF ESSENTIAL CALCULUS SKILLS

Example 27. Evaluate the following derivative with respect to x.

$$\frac{d}{dx}(6x^5)$$

Solution. When taking a derivative of a polynomial expression of the form $y = ax^b$ (where a and b are constants) with respect to x, the exponent (b) comes down to multiply the coefficient (a) and the exponent is reduced by 1 according to the following formula:

$$\frac{dy}{dx} = \frac{d}{dx}(ax^b) = bax^{b-1}$$

Compare $6x^5$ with ax^b to see that the coefficient is $a = 6$ and the exponent is $b = 5$. Plug these values into the previous formula.

$$\frac{d}{dx}(6x^5) = (5)(6)x^{5-1} = 30x^4$$

The answer is $30x^4$.

Example 28. Evaluate the following derivative with respect to t.

$$\frac{d}{dt}(t)$$

Solution. First note that this derivative is with respect to t (there are no x's in this problem). A derivative of an expression of the form $y = at^b$ is:

$$\frac{d}{dt}(at^b) = bat^{b-1}$$

Next, recall from algebra that a coefficient and exponent of 1 are always implied when you don't see them. This means that we can write t as $1t^1$. Compare $1t^1$ with at^b to see that $a = 1$ and $b = 1$. Now plug these values into the formula for the derivative.

$$\frac{d}{dt}(1t^1) = (1)(1)t^{1-1} = 1t^0 = 1$$

Recall the rule that $t^0 = 1$. The final answer is 1. Note that another way to write $\frac{d}{dt}(t)$ is $\frac{dt}{dt}$, where it may be more readily apparent that the answer is 1.

Example 29. Evaluate the following derivative with respect to x.

$$\frac{d}{dx}(2x^{-3})$$

Solution. Compare $2x^{-3}$ with ax^b to see that $a = 2$ and $b = -3$. Plug these into the formula for the derivative of a polynomial expression.

$$\frac{d}{dx}(2x^{-3}) = (-3)(2)x^{-3-1} = -6x^{-4}$$

Note that $-3 - 1 = -4$ (the exponent is reduced from -3 to -4). The answer is $-6x^{-4}$. Note that the answer can also be expressed as $-\frac{6}{x^4}$ since $x^{-4} = \frac{1}{x^4}$.

Example 30. Evaluate the following derivative with respect to t.

$$\frac{d}{dt}\left(\frac{3}{t^6}\right)$$

Solution. First apply the rule from algebra that $t^{-m} = \frac{1}{t^m}$.

$$\frac{d}{dt}\left(\frac{3}{t^6}\right) = \frac{d}{dt}(3t^{-6})$$

Next compare $3t^{-6}$ with at^b to see that $a = 3$ and $b = -6$. Now plug these values into the formula for the derivative.

$$\frac{d}{dt}(3t^{-6}) = (-6)(3)t^{-6-1} = -18t^{-7}$$

Note that $-6 - 1 = -7$ (the exponent is reduced from -6 to -7). The answer is $-18t^{-7}$.

Note that the answer can also be expressed as $-\frac{18}{t^7}$ since $x^{-7} = \frac{1}{x^7}$.

Example 31. Evaluate the following derivative with respect to x.

$$\frac{d}{dx}(3x^8 - 6x^5 + 9x^2 - 4)$$

Solution. Apply the following rule from calculus:

$$\frac{d}{dx}(y_1 + y_2 + y_3 + y_4) = \frac{dy_1}{dx} + \frac{dy_2}{dx} + \frac{dy_3}{dx} + \frac{dy_4}{dx}$$

This means that we can simply take a derivative of each term and then add the separate derivatives together.

$$\frac{d}{dx}(3x^8 - 6x^5 + 9x^2 - 4) = \frac{d}{dx}(3x^8) + \frac{d}{dx}(-6x^5) + \frac{d}{dx}(9x^2) + \frac{d}{dx}(-4)$$

Apply the formula $\frac{d}{dx}(ax^b) = bax^{b-1}$ to each term.

- The first term is $3x^8$. Compare $3x^8$ with ax^b to see that $a = 3$ and $b = 8$.

$$\frac{d}{dx}(3x^8) = (8)(3)x^{8-1} = 24x^7$$

- The second term is $-6x^5$. Compare $-6x^5$ with ax^b to see that $a = -6$ and $b = 5$.

$$\frac{d}{dx}(-6x^5) = (5)(-6)x^{5-1} = -30x^4$$

- The third term is $9x^2$. Compare $9x^2$ with ax^b to see that $a = 9$ and $b = 2$.

$$\frac{d}{dx}(9x^2) = (2)(9)x^{2-1} = 18x^1 = 18x$$

- The last term is -4. Recall from calculus that the derivative of a constant is zero.

$$\frac{d}{dx}(-4) = 0$$

Add these derivatives together.

$$\frac{d}{dx}(3x^8 - 6x^5 + 9x^2 - 4) = 24x^7 - 30x^4 + 18x + 0 = 24x^7 - 30x^4 + 18x$$

The answer is $24x^7 - 30x^4 + 18x$.

Example 32. Evaluate the following derivative with respect to u.

$$\frac{d}{du}\left(\frac{1}{\sqrt{u}}\right)$$

Solution. First apply the rule from algebra that $\frac{1}{\sqrt{u}} = u^{-1/2}$.

$$\frac{d}{du}\left(\frac{1}{\sqrt{u}}\right) = \frac{d}{du}\left(u^{-1/2}\right)$$

Next compare $u^{-1/2}$ with au^b to see that $a = 1$ (since a coefficient of 1 is implied when you don't see it) and $b = -\frac{1}{2}$. Now plug these values into the formula for the derivative.

$$\frac{d}{du}\left(u^{-1/2}\right) = -\frac{1}{2}u^{-1/2-1} = -\frac{1}{2}u^{-3/2}$$

In the last step, we subtracted 1 from the fraction by finding a common denominator: $-\frac{1}{2} - 1 = -\frac{1}{2} - \frac{2}{2} = \frac{-1-2}{2} = -\frac{3}{2}$. Note that this answer isn't in **standard form**. We must **rationalize the denominator** in order to express the answer in standard form. Apply the rule from algebra that $u^{-m} = \frac{1}{u^m}$.

$$-\frac{1}{2}u^{-3/2} = -\frac{1}{2u^{3/2}}$$

Multiply the numerator and denominator both by $u^{1/2}$.

$$-\frac{1}{2u^{3/2}} = -\frac{1}{2u^{3/2}}\frac{u^{1/2}}{u^{1/2}} = -\frac{u^{1/2}}{2u^{3/2}u^{1/2}}$$

Apply the rule from algebra that $x^m x^n = x^{m+n}$. When multiplying powers of the same base, add the exponents together.

$$-\frac{u^{1/2}}{2u^{3/2}u^{1/2}} = -\frac{u^{1/2}}{2u^{3/2+1/2}} = -\frac{u^{1/2}}{2u^2}$$

In the last line, we added the fractions: $\frac{3}{2} + \frac{1}{2} = \frac{3+1}{2} = \frac{4}{2} = 2$. Apply the rule $u^{1/2} = \sqrt{u}$.

$$-\frac{u^{1/2}}{2u^2} = -\frac{\sqrt{u}}{2u^2}$$

The final answer is $-\frac{\sqrt{u}}{2u^2}$. Note that the answer is the same as $-\frac{u^{1/2}}{2u^2}$, and the answer is also equivalent to $-\frac{1}{2}u^{-3/2}$ or $-\frac{1}{2u^{3/2}}$.

Example 33. Perform the following definite integral.

$$\int_{x=0}^{2} 8x^3 \, dx$$

Solution. Begin by finding the anti-derivative of the expression in the integrand: $8x^3$. When taking an anti-derivative of a polynomial expression of the form $y = ax^b$ (where a and b are constants, and where $b \neq -1$) over the variable x, the exponent (b) is raised by one power (from b to $b+1$) and the new exponent ($b+1$) comes down to the denominator according to the following formula:

$$\int y \, dx = \int ax^b \, dx = \frac{ax^{b+1}}{b+1} + c \qquad (b \neq -1)$$

The constant of integration (c) won't matter for the definite integral because it will cancel out (as we will see later). Compare $8x^3$ with ax^b to see that the coefficient is $a = 8$ and the exponent is $b = 3$. Plug these values into the previous formula.

$$\int 8x^3 \, dx = \frac{8x^{3+1}}{3+1} + c = \frac{8x^4}{4} + c = 2x^4 + c$$

The anti-derivative is $2x^4 + c$, but that's the indefinite integral. For the definite integral given in the problem, we must evaluate the anti-derivative function over the limits.

$$\int_{x=0}^{2} 8x^3 \, dx = [2x^4 + c]_{x=0}^{2}$$

The notation $[2x^4 + c]_{x=0}^{2}$ means to evaluate the function $2x^4 + c$ when $x = 2$, next evaluate the function $2x^4 + c$ when $x = 0$, and then to subtract the two results.

$$\int_{x=0}^{2} 8x^3 \, dx = [2x^4 + c]_{x=0}^{2} = [2(2)^4 + c] - [2(0)^4 + c] = 32 + c - 0 - c = 32$$

As expected, the constant of integration (c) canceled out in the definite integral. The final answer is 32.

Example 34. Perform the following definite integral.

$$\int_{t=-5}^{5} t^4 \, dt$$

Solution. First note that this integral is over t (there are no x's in this problem). An anti-derivative of an expression of the form $y = at^b$ is:

$$\int y \, dt = \int at^b \, dt = \frac{at^{b+1}}{b+1} + c \qquad (b \neq -1)$$

Next, recall from algebra that a coefficient of 1 is always implied when you don't see one. This means that we can write t^4 as $1t^4$. Compare $1t^4$ with at^b to see that $a = 1$ and $b = 4$. Now plug these values into the formula for the anti-derivative.

$$\int t^4 \, dt = \frac{t^{4+1}}{4+1} + c = \frac{t^5}{5} + c$$

This problem involves a definite integral, not an indefinite integral, so we need more than just the anti-derivative. We must evaluate the anti-derivative over the limits. We will ignore the constant of integration (c) since it will cancel out when we evaluate the definite integral over its limits (just like it did in the previous example).

$$\int_{t=-5}^{5} t^4 \, dt = \left[\frac{t^5}{5} \right]_{t=-5}^{5}$$

The notation $\left[\frac{t^5}{5} \right]_{t=-5}^{5}$ means to evaluate the function $\frac{t^5}{5}$ when $t = 5$, next evaluate the function $\frac{t^5}{5}$ when $t = -5$, and then to subtract the two results.

$$\int_{t=-5}^{5} t^4 \, dt = \left[\frac{t^5}{5} \right]_{t=-5}^{5} = \frac{(5)^5}{5} - \frac{(-5)^5}{5} = 625 - (-625) = 625 + 625 = 1250$$

Note that $\frac{(-5)^5}{5} = -\frac{3125}{5} = -625$ is negative, and subtracting a negative number equates to adding a positive number: $625 - (-625) = 625 + 625$. The final answer is 1250.

Example 35. Perform the following definite integral.

$$\int_{x=3}^{5} 900 \, x^{-3} \, dx$$

Solution. Compare $900x^{-3}$ with ax^b to see that $a = 900$ and $b = -3$. Plug these into the formula for the anti-derivative of a polynomial expression, and evaluate the result over the limits in order to perform the definite integral.

$$\int_{x=3}^{5} 900 \, x^{-3} \, dx = \left[\frac{ax^{b+1}}{b+1} \right]_{x=3}^{5} = \left[\frac{900x^{-3+1}}{-3+1} \right]_{x=3}^{5} = \left[\frac{900x^{-2}}{-2} \right]_{x=3}^{5} = [-450x^{-2}]_{x=3}^{5}$$

$$= \left[-\frac{450}{x^2} \right]_{x=3}^{5} = -\frac{450}{5^2} - \left(-\frac{450}{3^2} \right) = -18 - (-50) = -18 + 50 = 32$$

Note that $-3 + 1 = -2$. Going from the first line to the second line, we applied the rule from algebra that $x^{-2} = \frac{1}{x^2}$. Note that subtracting a negative number equates to adding a positive number: $-18 - (-50) = -18 + 50$. The final answer is 32.

Example 36. Perform the following definite integral.

$$\int_{u=1}^{3} \frac{81 \, du}{u^4}$$

Solution. First apply the rule from algebra that $\frac{1}{u^m} = u^{-m}$.

$$\int_{u=1}^{3} \frac{81 \, du}{u^4} = \int_{u=1}^{3} 81u^{-4} \, du$$

Compare $81u^{-4}$ with au^b to see that $a = 81$ and $b = -4$. Plug these into the formula for the anti-derivative of a polynomial expression, and evaluate the result over the limits in order to perform the definite integral.

$$\int_{u=1}^{3} 81u^{-4} \, du = \left[\frac{au^{b+1}}{b+1} \right]_{u=1}^{3} = \left[\frac{81u^{-4+1}}{-4+1} \right]_{u=1}^{3} = \left[\frac{81u^{-3}}{-3} \right]_{u=1}^{3} = [-27u^{-3}]_{u=1}^{3}$$

$$= \left[-\frac{27}{u^3} \right]_{u=1}^{3} = -\frac{27}{3^3} - \left(-\frac{27}{1^3} \right) = -\frac{27}{27} - \left(-\frac{27}{1} \right) = -1 - (-27) = -1 + 27 = 26$$

Note that $-4 + 1 = -3$. Going from the first line to the second line, we applied the rule from algebra that $u^{-3} = \frac{1}{u^3}$. Note that subtracting a negative number equates to adding a positive number: $-1 - (-27) = -1 + 27$. The final answer is 26.

Example 37. Perform the following definite integral.

$$\int_{x=4}^{9} \frac{3 \, dx}{\sqrt{x}}$$

Solution. Recall from algebra that $\sqrt{x} = x^{1/2}$ and $\frac{1}{x} = x^{-1}$ so that we can write $\frac{1}{\sqrt{x}} = x^{-1/2}$.

$$\int_{x=4}^{9} \frac{3 \, dx}{\sqrt{x}} = \int_{x=4}^{9} 3x^{-1/2} \, dx$$

Compare $3x^{-1/2}$ with ax^b to see that $a = 3$ and $b = -\frac{1}{2}$. Plug these into the formula for the anti-derivative of a polynomial expression, and evaluate the result over the limits in order to perform the definite integral.

$$\int_{x=4}^{9} 3x^{-1/2} \, dx = \left[\frac{au^{b+1}}{b+1} \right]_{x=4}^{9} = \left[\frac{3x^{-1/2+1}}{-1/2+1} \right]_{x=4}^{9} = \left[\frac{3x^{1/2}}{1/2} \right]_{x=4}^{9} = \left[6x^{1/2} \right]_{x=4}^{9}$$

$$= \left[6\sqrt{x} \right]_{x=4}^{9} = 6\sqrt{9} - 6\sqrt{4} = 6(3) - 6(2) = 18 - 12 = 6$$

Note that $-\frac{1}{2} + 1 = \frac{1}{2}$ and $\frac{3}{1/2} = 3(2) = 6$ (to divide by a fraction, multiply by its **reciprocal**).

Going from the first line to the second line, we applied the rule from algebra that $x^{1/2} = \sqrt{x}$. The final answer is 6.

Example 38. Perform the following definite integral.

$$\int\limits_{t=1}^{16} 14\, t^{3/4}\, dt$$

Solution. Compare $14\, t^{3/4}$ with at^b to see that $a = 14$ and $b = \frac{3}{4}$. Plug these into the formula for the anti-derivative of a polynomial expression, and evaluate the result over the limits in order to perform the definite integral.

$$\int\limits_{t=1}^{16} 14\, t^{3/4}\, dt = \left[\frac{at^{b+1}}{b+1}\right]_{t=1}^{16} = \left[\frac{14t^{3/4+1}}{3/4+1}\right]_{t=1}^{16} = \left[\frac{14t^{7/4}}{7/4}\right]_{t=1}^{16} = \left[8t^{7/4}\right]_{t=1}^{16}$$

$$= 8(16)^{7/4} - 8(1)^{7/4} = 8\left(\sqrt[4]{16}\right)^7 - 8\left(\sqrt[4]{1}\right)^7 = 8(2)^7 - 8(1)^7$$

$$= 8(128) - 8(1) = 1024 - 8 = 1016$$

Note that $\frac{3}{4} + 1 = \frac{3}{4} + \frac{4}{4} = \frac{7}{4}$ (add fractions with a **common denominator**) and $\frac{14}{7/4} = \frac{14(4)}{7} = 8$ (to divide by a fraction, multiply by its **reciprocal**). Also note that $\sqrt[4]{16}$ is the fourth root of 16: We wrote $16^{7/4}$ as $\left(\sqrt[4]{16}\right)^7$, though you could use a calculator to see that $16^{7/4} = 128$. Enter this as 16^(7/4) using parentheses. It's 16 raised to the power of $\frac{7}{4}$ (it is **not** 16 times $\frac{7}{4}$.) The final answer is 1016.

Example 39. Perform the following definite integral.

$$\int\limits_{x=2}^{4} (12x^3 - 9x^2)\, dx$$

Solution. Apply the following rule from calculus:

$$\int (y_1 + y_2)\, dx = \int y_1\, dx + \int y_2\, dx$$

This means that we can integrate over each term separately:

$$\int\limits_{x=2}^{4} (12x^3 - 9x^2)\, dx = \int\limits_{x=2}^{4} 12x^3\, dx + \int\limits_{x=2}^{4} -9x^2\, dx = \int\limits_{x=2}^{4} 12x^3\, dx - \int\limits_{x=2}^{4} 9x^2\, dx$$

$$= \left[\frac{12x^{3+1}}{3+1}\right]_{x=2}^{4} - \left[\frac{9x^{2+1}}{2+1}\right]_{x=2}^{4} = \left[\frac{12x^4}{4}\right]_{x=2}^{4} - \left[\frac{9x^3}{3}\right]_{x=2}^{4} = [3x^4]_{x=2}^{4} - [3x^3]_{x=2}^{4}$$

$$= [3(4)^4 - 3(2)^4] - [3(4)^3 - 3(2)^3] = (768 - 48) - (192 - 24) = 720 - 168 = 552$$

The answer is 552.

Example 40. Perform the following definite integral.

$$\int_{x=-1}^{2} (12x^5 + 24x^3 - 48x)\, dx$$

Solution. Apply the following rule from calculus:

$$\int (y_1 + y_2 + y_3)\, dx = \int y_1\, dx + \int y_2\, dx + \int y_3\, dx$$

This means that we can integrate over each term separately:

$$\int_{x=-1}^{2} (12x^5 + 24x^3 - 48x)\, dx = \int_{x=-1}^{2} 12x^5\, dx + \int_{x=-1}^{2} 24x^3\, dx + \int_{x=-1}^{2} -48x\, dx$$

$$= \int_{x=-1}^{2} 12x^5\, dx + \int_{x=-1}^{2} 24x^3\, dx - \int_{x=-1}^{2} 48x\, dx$$

$$= \left[\frac{12x^{5+1}}{5+1}\right]_{x=-1}^{2} + \left[\frac{24x^{3+1}}{3+1}\right]_{x=-1}^{2} - \left[\frac{48x^{1+1}}{1+1}\right]_{x=-1}^{2}$$

$$= \left[\frac{12x^6}{6}\right]_{x=-1}^{2} + \left[\frac{24x^4}{4}\right]_{x=-1}^{2} - \left[\frac{48x^2}{2}\right]_{x=-1}^{2}$$

$$= [2x^6]_{x=-1}^{2} + [6x^4]_{x=-1}^{2} - [24x^2]_{x=-1}^{2}$$

$$= [2(2)^6 - 2(-1)^6] + [6(2)^4 - 6(-1)^4] - [24(2)^2 - 24(-1)^2]$$

$$= (128 - 2) + (96 - 6) - (96 - 24) = 126 + 90 - 72 = 144$$

The answer is 144.

3 ONE-DIMENSIONAL UNIFORM ACCELERATION

Equations	Symbol	Name	SI Units
$\Delta x = v_{x0}t + \dfrac{1}{2}a_x t^2$ $v_x = v_{x0} + a_x t$ $v_x^2 = v_{x0}^2 + 2a_x \Delta x$	Δx	net displacement	m
	v_{x0}	initial velocity	m/s
	v_x	final velocity	m/s
	a_x	acceleration	m/s^2
	t	time	s

Example 41. A monkey drives a bananamobile with uniform acceleration. Starting from rest, the bananamobile travels 90 m in a time of 6.0 s. What is the acceleration of the bananamobile?

Solution. Begin with a labeled diagram. The car drives in a straight line. We choose $+x$ to point in the forward direction. The points i and f mark the initial and final positions.

$$i \longrightarrow f$$
$$+x$$

The unknown we are looking for is acceleration (a_x). List the three knowns.
- The net displacement is $\Delta x = 90$ m.
- The time is $t = 6.0$ s.
- The initial velocity is $v_{x0} = 0$ because the bananamobile begins from **rest**.

Since we know Δx, t, and v_{x0}, and since we're solving for a_x, we should use an equation that only has these four symbols. That would be the first equation of uniform acceleration:

$$\Delta x = v_{x0}t + \frac{1}{2}a_x t^2$$

Plug the knowns into this equation. To avoid clutter, suppress the units until the end.

$$90 = 0(6) + \frac{1}{2}a_x(6)^2$$

Note that $0(6) = 0$ and $\frac{1}{2}a_x(6)^2 = \frac{1}{2}a_x 36 = 18a_x$.

$$90 = 18a_x$$

To solve for the acceleration, divide both sides of the equation by 18.

$$a_x = 5.0 \text{ m/s}^2$$

The answer is $a_x = 5.0$ m/s^2.

Example 42. A mechanical monkey toy has an initial speed of 15 m/s, has uniform acceleration of -4.0 m/s^2, and travels for 6.0 seconds. What is the final velocity of the mechanical monkey?

Solution. Begin with a labeled diagram. The toy travels in a straight line. We choose $+x$ to point in the forward direction. The points i and f mark the initial and final positions. (As shown in the diagram, the toy actually runs out of speed and then travels backwards before the 6.0 s is up. However, it's not necessary to realize this in order to solve the problem.)

The unknown we are looking for is final velocity (v_x). List the three knowns.
- The initial velocity is $v_{x0} = 15$ m/s.
- The acceleration is $a_x = -4.0$ m/s^2.
- The time is $t = 6.0$ s.

Since we know v_{x0}, a_x, and t, and since we're solving for v_x, we should use an equation that only has these four symbols. That would be the second equation of uniform acceleration.
$$v_x = v_{x0} + a_x t$$
Plug the knowns into this equation. To avoid clutter, suppress the units until the end.
$$v_x = 15 + (-4)(6)$$
$$v_x = 15 - 24$$
$$v_x = -9.0 \text{ m/s}$$
The answer is $v_x = -9.0$ m/s. The significance of the minus sign is that the toy is traveling backward in the final position.

Notes: It's instructive to note that neither the initial velocity (v_{x0}) nor the final velocity (v_x) is zero in this problem. The initial (i) and final (f) points refer to positions where we know information or where we're solving for information: They do **not** necessarily refer to points where the motion began or where the motion ended.

Example 43. A monkey drives a bananamobile with uniform acceleration, beginning with a speed of 10 m/s and ending with a speed of 30 m/s. The acceleration is 8.0 m/s^2. How far does the monkey travel during this time?

Solution. Begin with a labeled diagram. The car drives in a straight line. We choose $+x$ to point in the forward direction. The points i and f mark the initial and final positions.

The unknown we are looking for is net displacement (Δx). List the three knowns.
- The initial velocity is $v_{x0} = 10$ m/s.
- The final velocity is $v_x = 30$ m/s.
- The acceleration is $a_x = 8.0$ m/s^2.

Since we know v_{x0}, v_x, and a_x, and since we're solving for Δx, we should use an equation that only has these four symbols. That would be the third equation of uniform acceleration.

$$v_x^2 = v_{x0}^2 + 2a_x\Delta x$$

Plug the knowns into this equation. To avoid clutter, suppress the units until the end.

$$(30)^2 = (10)^2 + 2(8)\Delta x$$

Simplify this equation.

$$900 = 100 + 16\Delta x$$

Subtract 100 from both sides of the equation to isolate the unknown term.

$$800 = 16\Delta x$$

To solve for the net displacement, divide both sides of the equation by 16.

$$\Delta x = 50 \text{ m}$$

The answer is $\Delta x = 50$ m.

Equations	Symbol	Name	SI Units
	Δy	net displacement	m
$\Delta y = v_{y0}t + \dfrac{1}{2}a_y t^2$	v_{y0}	initial velocity	m/s
$v_y = v_{y0} + a_y t$	v_y	final velocity	m/s
$v_y^2 = v_{y0}^2 + 2a_y\Delta y$	a_y	acceleration	m/s^2
	t	time	s

Note: It's customary to use x for **horizontal** motion and to use y for **vertical** motion.

Example 44. On Planet Fyzx, a chimpanzee astronaut drops a 500-g banana from rest from a height of 36 m above the ground and the banana strikes the ground 4.0 s later. What is the acceleration of the banana?

Solution. Begin with a labeled diagram, including the path and the initial (i) and final (f) positions. Choose the $+y$-direction to be upward even though the banana falls downward. This choice makes it easier to reason out the signs correctly.

Ignore the 500 g. The mass of the banana **doesn't** affect the answer. (To see this, try dropping two rocks with different masses from the same height at the same time. You should observe that the rocks strike the ground at about the same time, regardless of which is heavier.) Note that we will neglect the effects of air resistance unless stated otherwise. Since this is a **free fall** problem, we will apply the equations of **uniform acceleration**.

The unknown we are looking for is acceleration (a_y). List the three knowns.

- The net displacement is $\Delta y = -36$ m. It's negative because the final position (f) is **below** the initial position (i).
- The time is $t = 4.0$ s.
- The initial velocity is $v_{y0} = 0$ because the banana is released from **rest**.

Since we know Δx, t, and v_{y0}, and since we're solving for a_y, we should use an equation that only has these four symbols. That would be the first equation of uniform acceleration.

$$\Delta y = v_{y0}t + \frac{1}{2}a_yt^2$$

Plug the knowns into this equation. To avoid clutter, suppress the units until the end.

$$-36 = 0(4) + \frac{1}{2}a_y(4)^2$$

Note that $0(4) = 0$ and $\frac{1}{2}a_y(4)^2 = \frac{1}{2}a_y16 = 8a_y$.

$$-36 = 8a_y$$

To solve for the acceleration, divide both sides of the equation by 8.

$$a_y = -\frac{36}{8} = -\frac{9}{2} \text{ m/s}^2 = -4.5 \text{ m/s}^2$$

Note that a_y is negative because the force of gravity is downward (and we chose the $+y$-direction to be upward). The answer is $a_y = -\frac{9}{2}$ m/s$^2 = -4.5$ m/s^2.

Example 45. A monkey leans over the edge of a cliff and throws a banana straight upward with a speed of 20 m/s. The banana lands on the ground, 60 m below its starting position. For how much time is the banana in the air?

Solution. Begin with a labeled diagram, including the path and the initial (i) and final (f) positions. Choose the $+y$-direction to be upward (so it's easier to reason out the signs).

The unknown we are looking for is time (t). List the three knowns.

- The initial velocity is $v_{y0} = 20$ m/s.
- The net displacement is $\Delta y = -60$ m. It's negative because the final position (f) is **below** the initial position (i).
- The acceleration is $a_y = -9.81$ m/s^2 because the banana is in free fall near earth's surface. (Assume all problems are near earth's surface unless stated otherwise.) Note that a_y is negative for free fall problems (since we chose $+y$ to be up).

As usual, we neglect air resistance unless the problem states otherwise. Since we know v_{y0}, Δy, and a_y, and since we're solving for t, we should use an equation that only has these four symbols. That would be the first equation of uniform acceleration.

$$\Delta y = v_{y0}t + \frac{1}{2}a_y t^2$$

Plug the knowns into this equation. To avoid clutter, suppress the units until the end.

$$-60 = 20t + \frac{1}{2}(-9.81)t^2$$

Simplify this equation.

$$-60 = 20t - 4.905t^2$$

Recognize that this is a **quadratic equation** because it includes a quadratic term ($-4.905t^2$), a linear term ($20t$), and a constant term (-60). Use algebra to bring $-4.905t^2$ and $20t$ to the left-hand side, so that all three terms are on the same side of the equation. (These terms will change sign when we add $4.905t^2$ to both sides and subtract $20t$ from both sides.) Also, order the terms such that the equation is in **standard form**, with the quadratic term first, the linear term second, and the constant term last.

$$4.905t^2 - 20t - 60 = 0$$

Compare this equation to the general form $at^2 + bt + c = 0$ to identify the constants.

$$a = 4.905 \quad , \quad b = -20 \quad , \quad c = -60$$

Plug these constants into the **quadratic formula**.

$$t = \frac{-b \pm \sqrt{b^2 - 4ac}}{2a} = \frac{-(-20) \pm \sqrt{(-20)^2 - 4(4.905)(-60)}}{2(4.905)}$$

Note that $-(-20) = +20$ (two minus signs make a plus sign), $(-20)^2 = +400$ (it's positive since the minus sign is squared), and $-4(4.905)(-60) = +1177.2$ (two minus signs make a plus sign).

$$t = \frac{20 \pm \sqrt{400 + 1177.2}}{9.81} = \frac{20 \pm \sqrt{1577.2}}{9.81} = \frac{20 \pm 39.71}{9.81}$$

We must consider both solutions. Work out the two cases separately.

$$t = \frac{20 + 39.71}{9.81} \quad \text{or} \quad t = \frac{20 - 39.71}{9.81}$$

$$t = \frac{59.71}{9.81} \quad \text{or} \quad t = \frac{-19.71}{9.81}$$

$$t = 6.09 \text{ s} \quad \text{or} \quad t = -2.01 \text{ s}$$

Since time can't be negative, the correct answer is $t = 6.09$ s. (Note that if you round 9.81 to 10, you can get the answers ≈ 6.0 s and ≈ -2.0 s without using a calculator.)

Example 46. Monk Jordan leaps straight upward. He spends exactly 1.00 second in the air. With what initial speed does Monk Jordan jump?

Solution. Begin with a labeled diagram, including the path and the initial (i) and final (f) positions. Choose the $+y$-direction to be upward (so it's easier to reason out the signs).

One way to solve this problem is to work with only **half the trip** (that's why we labeled the final position at the top of the trajectory in our diagram). The unknown we are looking for is initial velocity (v_{y0}). List the three knowns.

- The time is $t = 0.50$ s (since we are working with half the trip, not the whole trip).
- The final velocity is $v_y = 0$ (since we put the final position at the top of the path).
- The acceleration is $a_y = -9.81$ m/s^2 because the banana is in free fall near earth's surface. Note that a_y is negative since we chose $+y$ to be upward.

Since we know t, v_y, and a_y, and since we're solving for v_{y0}, we should use an equation that only has these four symbols. That would be the second equation of uniform acceleration.

$$v_y = v_{y0} + a_y t$$

Plug the knowns into this equation. To avoid clutter, suppress the units until the end.

$$0 = v_{y0} + (-9.81)(0.50)$$

Simplify this equation.

$$0 = v_{y0} - 4.905$$

To solve for the initial velocity, add 4.905 to both sides of the equation.

$$v_{y0} = 4.9 \text{ m/s}$$

The answer is $v_{y0} = 4.9$ m/s (to two significant figures).

Alternate solution. If instead you choose to work with the roundtrip (rather than just the trip upward), the knowns would be:

- $\Delta y = 0$ since the final position would be at the same height as the initial position.
- $t = 1.00$ s for the whole trip (instead of just the trip up).
- $a_y = -9.81$ m/s^2 (the same for all free fall problems near earth's surface).

Using these knowns, you could then use the equation $\Delta y = v_{y0} t + \frac{1}{2} a_y t^2$ to solve for the initial velocity. You would obtain the same answer, $v_{y0} = 4.9$ m/s.

4 ONE-DIMENSIONAL MOTION WITH CALCULUS

Equations	Symbol	Name	SI Units
$v_x = \dfrac{dx}{dt}$	x	position coordinate	m
$a_x = \dfrac{dv_x}{dt} = \dfrac{d^2x}{dt^2}$	Δx	net displacement	m
	v_{x0}	initial velocity	m/s
$\Delta x = \displaystyle\int_{t=t_0}^{t} v_x \, dt$	v_x	velocity	m/s
	a_x	acceleration	m/s^2
$v_x = v_{x0} + \displaystyle\int_{t=t_0}^{t} a_x \, dt$	t_0	initial time	s
	t	time	s

Example 47. The position of a bananamobile is given according to the following equation, where SI units have been suppressed.

$$x = 4t^3 - 8t^2 + 64$$

(A) Determine the velocity of the bananamobile at $t = 5.0$ s.

Solution. Velocity is the first **derivative** of position with respect to time.

$$v_x = \frac{dx}{dt} = \frac{d}{dt}(4t^3 - 8t^2 + 64) = \frac{d}{dt}(4t^3) + \frac{d}{dt}(-8t^2) + \frac{d}{dt}(64)$$

Recall the formula for the derivative of a polynomial expression: $\frac{d}{dt}(at^b) = bat^{b-1}$.

$$v_x = \frac{dx}{dt} = (3)(4)t^{3-1} - (2)(8)t^{2-1} + 0 = 12t^2 - 16t$$

Note that the derivative of a constant equals zero: $\frac{d}{dt}(64) = 0$. Evaluate the velocity function at the specified time ($t = 5.0$ s).

$$v_x(\text{at } t = 5 \text{ s}) = 12(5)^2 - 16(5) = 12(25) - 80 = 300 - 80 = 220 \text{ m/s}$$

The velocity at the specified time is $v_x = 220$ m/s.

(B) Determine the acceleration of the bananamobile at $t = 7.0$ s.

Solution. Acceleration is the first **derivative** of velocity with respect to time. Be careful to take a derivative of the velocity **function** ($12t^2 - 16t$) that we found in part (A), **not** the value of 220 m/s found after plugging in time.

$$a_x = \frac{dv_x}{dt} = \frac{d}{dt}(12t^2 - 16t) = \frac{d}{dt}(12t^2) + \frac{d}{dt}(-16t)$$

$$a_x = \frac{dv_x}{dt} = (2)(12)t^{2-1} - (1)(16)t^0 = 24t - 16$$

Note that $\frac{d}{dt}(-16t) = -16$. Also recall from algebra that $t^0 = 1$. Evaluate the acceleration function at the specified time ($t = 7.0$ s). Note that the time specified in part (B) is **not** the same as the time that was specified in part (A).

$$a_x(\text{at } t = 7 \text{ s}) = 24(7) - 16 = 168 - 16 = 152 \text{ m/s}^2$$

The acceleration at the specified time is $a_x = 152$ m/s^2.

Example 48. As you begin your physics exam, a monkey grabs your pencil and runs according to the following equation, where SI units have been suppressed.

$$v_x = \sqrt{2t}$$

(A) Determine the net displacement of the monkey from $t = 2.0$ s until $t = 8.0$ s.

Solution. Net displacement is the definite **integral** of velocity over time.

$$\Delta x = \int_{t=t_0}^{t} v_x \, dt = \int_{t=2}^{8} \sqrt{2t} \, dt$$

First apply the rules from algebra that $\sqrt{ax} = \sqrt{a}\sqrt{x}$ and $x^{1/2} = \sqrt{x}$ to write $\sqrt{2t}$ as $2^{1/2}t^{1/2}$.

$$\Delta x = \int_{t=2}^{8} 2^{1/2}t^{1/2} \, dt$$

Compare $2^{1/2}t^{1/2}$ with at^b to see that $a = 2^{1/2}$ and $b = \frac{1}{2}$. Plug these into the formula for the anti-derivative of a polynomial expression, and evaluate the result over the limits in order to perform the definite integral.

$$\Delta x = \left[\frac{2^{1/2}t^{1/2+1}}{\frac{1}{2}+1}\right]_{t=2}^{8} = \left[\frac{2^{1/2}t^{3/2}}{\frac{3}{2}}\right]_{t=2}^{8} = \left[\frac{2}{3}2^{1/2}t^{3/2}\right]_{t=2}^{8} = \left[\frac{2^{1/2+1}t^{3/2}}{3}\right]_{t=2}^{8} = \left[\frac{2^{3/2}t^{3/2}}{3}\right]_{t=2}^{8}$$

Note that $\frac{1}{2} + 1 = \frac{1}{2} + \frac{2}{2} = \frac{3}{2}$ (add fractions with a **common denominator**) and $\frac{2^{1/2}}{3/2} = \frac{2(2^{1/2})}{3} = \frac{2^1 2^{1/2}}{3} = \frac{2^{1+1/2}}{3} = \frac{2^{3/2}}{3}$ (to divide by a fraction, multiply by its **reciprocal**).

$$\Delta x = \left[\frac{(2t)^{3/2}}{3}\right]_{t=2}^{8} = \frac{[2(8)]^{3/2}}{3} - \frac{[2(2)]^{3/2}}{3} = \frac{[16]^{3/2}}{3} - \frac{[4]^{3/2}}{3}$$

Note that $[16]^{3/2} = \sqrt{16^3} = \left(\sqrt{16}\right)^3$, or use your calculator to see that 16^(3/2) equals 64 (be sure to use parentheses correctly if you try it). It's 16 raised to the power of $\frac{3}{2}$ (it is **not** 16 times $\frac{3}{2}$.)

$$\Delta x = \frac{\left(\sqrt{16}\right)^3}{3} - \frac{\left(\sqrt{4}\right)^3}{3} = \frac{4^3}{3} - \frac{2^3}{3} = \frac{64}{3} - \frac{8}{3} = \frac{64-8}{3} = \frac{56}{3} \text{ m}$$

The net displacement for the specified time interval is $\Delta x = \frac{56}{3}$ m.

(B) Determine the acceleration of the monkey at $t = 8.0$ s.

Solution. Acceleration is the first **derivative** of velocity with respect to time.

$$a_x = \frac{dv_x}{dt} = \frac{d}{dt}\left(\sqrt{2t}\right)$$

Apply the rules from algebra that $\sqrt{ax} = \sqrt{a}\sqrt{x}$ and $x^{1/2} = \sqrt{x}$ to write $\sqrt{2t}$ as $2^{1/2}t^{1/2}$.

$$a_x = \frac{d}{dt}\left(2^{1/2}t^{1/2}\right) = \frac{1}{2}2^{1/2}t^{1/2-1} = \frac{2^{1/2}}{2}t^{-1/2} = 2^{1/2-1}t^{-1/2} = 2^{-1/2}t^{-1/2}$$

Note that $\frac{1}{2} - 1 = \frac{1}{2} - \frac{2}{2} = -\frac{1}{2}$ (subtract fractions with a **common denominator**). Also note

that $\frac{2^{1/2}}{2} = \frac{2^{1/2}}{2^1} = 2^{1/2-1} = 2^{-1/2}$. Another way to look at this is that $\frac{2^{1/2}}{2}$ is the same thing as

$\frac{\sqrt{2}}{2}$ or $\frac{1}{\sqrt{2}}$ (which is the same as $2^{-1/2}$). Why does $\frac{\sqrt{2}}{2} = \frac{1}{\sqrt{2}}$? It's because $\sqrt{2}\sqrt{2} = 2$. If you

enter both $\frac{\sqrt{2}}{2}$ and $\frac{1}{\sqrt{2}}$ correctly on your calculator, you'll get the same decimal either way.

$$a_x = (2t)^{-1/2} = \frac{1}{(2t)^{1/2}} = \frac{1}{\sqrt{2t}}$$

We applied the rules $x^{-1/2} = \frac{1}{\sqrt{x}}$ and $\sqrt{ax} = \sqrt{a}\sqrt{x}$ to write $2^{1/2}t^{1/2}$ as $\frac{1}{\sqrt{2t}}$. Evaluate the acceleration function at the specified time ($t = 8.0$ s).

$$a_x(\text{at } t = 8 \text{ s}) = \frac{1}{\sqrt{2(8)}} = \frac{1}{\sqrt{16}} = \frac{1}{4} \text{ m/s}^2$$

The acceleration at the specified time is $a_x = \frac{1}{4}$ m/s^2.

Example 49. The acceleration of a chimpanzee is given according to the following equation, where SI units have been suppressed. The initial velocity of the chimpanzee is 18 m/s at $t = 0$.

$$a_x = 24t^2$$

(A) Determine the velocity of the chimpanzee at $t = 3.0$ s.

Solution. Velocity equals initial velocity plus the definite **integral** of acceleration over time.

$$v_x = v_{x0} + \int_{t=t_0}^{t} a_x\, dt = 18 + \int_{t=0}^{3} 24t^2\, dt$$

Compare $24t^2$ with at to see that $a = 24$ and $b = 2$. Plug these into the formula for the anti-derivative of a polynomial expression, and evaluate the result over the limits in order to perform the definite integral.

$$v_x = 18 + \left[\frac{24t^{2+1}}{2+1}\right]_{t=0}^{3} = 18 + \left[\frac{24t^3}{3}\right]_{t=0}^{3} = 18 + [8t^3]_{t=0}^{3}$$

$$v_x = 18 + 8(3)^3 - 8(0)^3 = 18 + 8(27) - 0 = 18 + 216 = 234 \text{ m/s}$$

The velocity at the specified time equals $v_x = 234$ m/s.

35

(B) Determine the net displacement of the chimpanzee from $t = 0$ until $t = 3.0$ s.

Solution. Net displacement is the definite **integral** of velocity over time. Be careful to integrate over the velocity **function** $(18 + 8t^3)$ that we found in part (A), **not** the value of 234 m/s found after plugging in time. (The function $18 + 8t^3$ was found in the form of the anti-derivative. The anti-derivative is a function of time, whereas the definite integral is a numerical value for specified time interval.)

$$\Delta x = \int_{t=t_0}^{t} v_x \, dt = \int_{t=0}^{3} (18 + 8t^3) \, dt$$

Apply the following rule from calculus:

$$\int (y_1 + y_2) \, dx = \int y_1 \, dx + \int y_2 \, dx$$

This means that we can integrate over each term separately:

$$\int_{t=0}^{3} (18 + 8t^3) \, dt = \int_{t=0}^{3} 18 \, dt + \int_{t=0}^{3} 8t^3 \, dt$$

Find the anti-derivative of each expression, and then evaluate the result over the limits in order to perform the definite integral. Note that the anti-derivative of a constant equals the constant times the integration variable: $\int 18 \, dt = 18t$. That's because $\frac{d}{dt}(18t) = 18$.

$$\Delta x = [18t]_{t=0}^{3} + \left[\frac{8t^{3+1}}{3+1}\right]_{t=0}^{3} = [18t]_{t=0}^{3} + \left[\frac{8t^4}{4}\right]_{t=0}^{3} = [18t]_{t=0}^{3} + [2t^4]_{t=0}^{3}$$

$$\Delta x = [18(3) - 18(0)] + [2(3)^4 - 2(0)^4] = 54 - 0 + 162 - 0 = 216 \text{ m}$$

The net displacement for the specified time interval is $\Delta x = 216$ m.

5 REVIEW OF ESSENTIAL GEOMETRY SKILLS

Example 50. Find the perimeter and area of the rectangle illustrated below.

4 m

6 m

Solution. Identify the width and length from the figure.
- The length of the rectangle is $L = 6$ m.
- The width of the rectangle is $W = 4$ m.

Plug the values for the length and width into the formula for perimeter.
$$P = 2L + 2W = 2(6) + 2(4) = 12 + 8 = 20 \text{ m}$$
Now plug the length and width into the formula for area.
$$A = LW = (6)(4) = 24 \text{ m}^2$$
The perimeter is $P = 20$ m and the area is $A = 24 \text{ m}^2$.

Example 51. Find the perimeter and area of the right triangle illustrated below.

3 m

4 m

Solution. Identify the base and height from the figure.
- The height of the triangle is $a = 3$ m.
- The base of the triangle is $b = 4$ m.

We need to find the hypotenuse of the triangle before we can find the perimeter. Use the Pythagorean theorem to solve for the hypotenuse.
$$a^2 + b^2 = c^2$$
$$3^2 + 4^2 = c^2$$
$$9 + 16 = 25 = c^2$$
Squareroot both sides of the equation to solve for the unknown.
$$c = \sqrt{25} = 5 \text{ m}$$
Add up the lengths of the sides to find the perimeter of the triangle.
$$P = a + b + c = 3 + 4 + 5 = 12 \text{ m}$$
Now plug the base and height into the formula for area.
$$A = \frac{1}{2}bh = \frac{1}{2}(4)(3) = 6 \text{ m}^2$$
The perimeter is $P = 12$ m and the area is $A = 6 \text{ m}^2$.

Example 52. If the area of a square is 36 m², what is its perimeter?

Solution. First solve for the length of an edge, given that the area is $A = 36$ m². Use the formula for the area of a square.

$$A = L^2$$
$$36 = L^2$$

Squareroot both sides of the equation to solve for the unknown.

$$L = \sqrt{36} = 6 \text{ m}$$

Set $W = L$ in the formula for the perimeter of a rectangle (since $W = L$ for a square). Plug the values for the length and width into the formula for perimeter.

$$P = 2L + 2W = 2L + 2L = 4L = 4(6) = 24 \text{ m}$$

The perimeter is $P = 24$ m. Alternatively, $P = 2L + 2W = 2(6) + 2(6) = 12 + 12 = 24$ m.

Example 53. Determine the hypotenuse of the right triangle illustrated below.

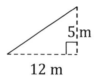

5 m

12 m

Solution. Identify the given sides of the triangle from the figure.
- The height of the triangle is $a = 5$ m.
- The base of the triangle is $b = 12$ m.

Use the Pythagorean theorem:

$$a^2 + b^2 = c^2$$
$$5^2 + 12^2 = c^2$$
$$25 + 144 = 169 = c^2$$

Squareroot both sides of the equation to solve for the unknown:

$$c = \sqrt{169} = 13 \text{ m}$$

The hypotenuse is $c = 13$ m.

Example 54. Determine the unknown side of the right triangle illustrated below.

2 m

$\sqrt{3}$ m

Solution. Identify the given sides of the triangle from the figure.
- The base of the triangle is $b = \sqrt{3}$ m.
- The hypotenuse of the triangle is $c = 2$ m.

Use the Pythagorean theorem:

$$a^2 + b^2 = c^2$$

38

$$a^2 + \left(\sqrt{3}\right)^2 = (2)^2$$

Recall from algebra that $\left(\sqrt{x}\right)^2 = \sqrt{x}\sqrt{x} = x$. Therefore, $\left(\sqrt{3}\right)^2 = 3$:

$$a^2 + 3 = 4$$

Subtract 3 from both sides of the equation to isolate the unknown term:

$$a^2 = 4 - 3 = 1$$

Squareroot both sides of the equation to solve for the unknown:

$$a = \sqrt{1} = 1 \text{ m}$$

The answer is $a = 1$ m.

Example 55. Determine the length of the diagonal of the rectangle illustrated below.

6 m

8 m

Solution. The diagonal divides the rectangle into two right triangles. Work with one of these right triangles. Identify the given sides of the triangle from the figure.

- The height of the triangle is $a = 6$ m.
- The base of the triangle is $b = 8$ m.

Use the Pythagorean theorem:

$$a^2 + b^2 = c^2$$
$$6^2 + 8^2 = c^2$$
$$36 + 64 = 100 = c^2$$

Squareroot both sides of the equation to solve for the unknown:

$$c = \sqrt{100} = 10 \text{ m}$$

The hypotenuse is $c = 10$ m.

Example 56. Find the radius, circumference, and area of the circle illustrated below.

6 m

Solution. The indicated diameter is $D = 6$ m. Use the equation for diameter to determine the radius.

$$D = 2R$$

To solve for the radius, divide both sides of the equation by 2.

$$R = \frac{D}{2} = \frac{6}{2} = 3 \text{ m}$$

Next plug the radius into the formula for circumference. Recall from math that $\pi \approx 3.14$.

$$C = 2\pi R = 2\pi(3) = 6\pi \approx 6(3.14) \approx 19 \text{ m}$$

Finally, plug the radius into the formula for area:
$$A = \pi R^2 = \pi(3)^2 = 9\pi \approx 9(3.14) \approx 28 \text{ m}^2$$
The answers are $R = 3$ m, $C = 6\pi$ m ≈ 19 m, and $A = 9\pi$ m$^2 \approx 28$ m^2.

Example 57. If the area of a circle is 16π m^2, what is its circumference?

Solution. First solve for the radius, given that the area is $A = 16\pi$ m^2. Use the equation for the area of a circle to solve for the radius.
$$A = \pi R^2$$
Divide both sides by π.
$$\frac{A}{\pi} = R^2$$
To solve for the radius, squareroot both sides of the equation.
$$R = \sqrt{\frac{A}{\pi}} = \sqrt{\frac{16\pi}{\pi}} = \sqrt{16} = 4 \text{ m}$$
Now plug the radius into the formula for circumference.
$$C = 2\pi R = 2\pi(4) = 8\pi \approx 8(3.14) \approx 25 \text{ m}$$
The circumference is $C = 8\pi$ m ≈ 25 m.

6 MOTION GRAPHS

How to Analyze a Motion Graph

For a graph of **position** (x) as a function of time (t):

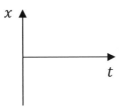

1. Just read the graph directly to find **net displacement** or **total distance traveled**. For the total distance traveled, add up each distance traveled forward or backward in absolute values (as in the first example that follows). For net displacement, use the formula $ND = x_f - x_i$, where x_f is the final position and x_i is the initial position.
2. Find the slope of the tangent line to find **velocity**.

For a graph of **velocity** (v_x) as a function of time (t):

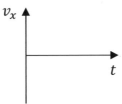

3. Find the area between the curve and the horizontal axis to determine the **net displacement** or the **total distance traveled**. For net displacement, any area below the horizontal axis is negative. For total distance traveled, all areas are positive.
4. Just read the graph directly to find **velocity**.
5. Find the slope of the tangent line to find **acceleration**.

For a graph of **acceleration** (a_x) as a function of time (t):

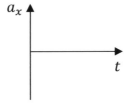

6. Find the area between the curve and the horizontal axis to determine the **change in velocity**. Any area below the horizontal axis is negative. Then use this equation:
$$v_x = v_{x0} + area$$
7. Just read the graph directly to find **acceleration**.

Example 58. A monkey drives a jeep. The position as a function of time is illustrated below.

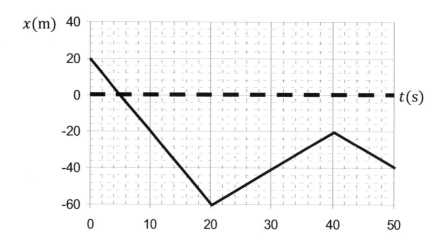

(A) Find the total distance traveled and net displacement for the trip.

Solution. To find net displacement from a position graph, read the initial and final values.

- The initial position is $x_i = 20$ m (read the vertical axis when $t = 0$).
- The final position is $x_f = -40$ m (read the vertical axis when $t = 50$ s).

Use the formula for net displacement.

$$ND = x_f - x_i = -40 - 20 = -60 \text{ m}$$

To find total distance traveled from a position graph, read the position values to determine how far the object moves forward or backward in each segment of the trip, and then add these increments in absolute values.

- For the first 20 s, the object moves backward 80 m (from $x = 20$ m to $x = -60$ m).
- For the next 20 s, the object moves forward 40 m (from $x = -60$ m to $x = -20$ m).
- For the last 10 s, the object moves backward 20 m (from $x = -20$ m to $x = -40$ m).

Add these increments up in absolute values.

$$TDT = |d_1| + |d_2| + |d_3| = |-80| + |40| + |-20| = 80 + 40 + 20 = 140 \text{ m}$$

The net displacement is $ND = -60$ m and the total distance traveled is $TDT = 140$ m.

(B) Find the velocity of the jeep at $t = 10$ s.

Solution. To find velocity from a position graph, find the slope. We want the slope of the line where $t = 10$ s. Read off the coordinates of the endpoints of this line segment.

$$(t_1, x_1) = (0, 20 \text{ m})$$
$$(t_2, x_2) = (20 \text{ s}, -60 \text{ m})$$

Note that position (x) is on the vertical axis, while time (t) is on the horizontal axis. Plug these values into the equation for slope.

$$v_x = \frac{x_2 - x_1}{t_2 - t_1} = \frac{-60 - 20}{20 - 0} = \frac{-80}{20} = -4.0 \text{ m/s}$$

The velocity is $v_x = -4.0$ m/s during the first 20 s. The velocity is negative because the slope is negative: The object is moving backward.

Example 59. A monkey rides a bike. The velocity as a function of time is illustrated below.

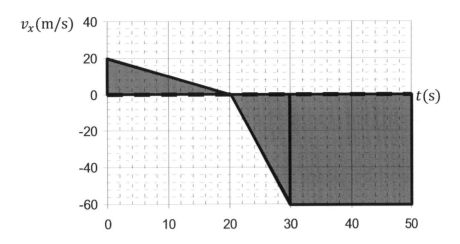

(A) What is the bike's acceleration at $t = 25$ seconds?

Solution. To find acceleration from a velocity graph, find the slope. We want the slope of the line where $t = 25$ s. Read off the coordinates of the endpoints:
$$(t_1, v_{1x}) = (20 \text{ s}, 0)$$
$$(t_2, v_{2x}) = (30 \text{ s}, -60 \text{ m/s})$$
Plug these values into the equation for slope:
$$a_x = \frac{v_{2x} - v_{1x}}{t_2 - t_1} = \frac{-60 - 0}{30 - 20} = \frac{60}{10} = -6.0 \text{ m/s}^2$$
Note that time (t) is on the horizontal axis, while velocity (v_x) is on the vertical axis. The acceleration is $a_x = -6.0$ m/s^2 during the second segment.

(B) What is the bike's net displacement for the whole trip?

Solution. To find net displacement from a velocity graph, divide the region between the curve (in this case the "curve" is made up of straight lines) and the horizontal axis into triangles and rectangles. See the two triangles and rectangle in the diagram above.
$$A_1 = \frac{1}{2} b_1 h_1 = \frac{1}{2}(20)(20) = 200 \text{ m}$$
$$A_2 = \frac{1}{2} b_2 h_2 = \frac{1}{2}(10)(-60) = -300 \text{ m}$$
$$A_3 = L_3 W_3 = (20)(-60) = -1200 \text{ m}$$
Note that the last two areas are negative because they lie below the t-axis. (However, if we had been finding total distance traveled, we would instead make all of the areas positive.) Add these three areas together:
$$ND = A_1 + A_2 + A_3 = 200 - 300 - 1200 = -1300 \text{ m}$$
The net displacement is $ND = -1300$ m $= -1.3$ km.

Example 60. A monkey flies a bananaplane. The acceleration as a function of time is illustrated below. The initial velocity of the bananaplane is 150 m/s.

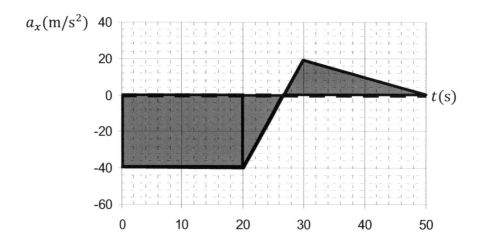

(A) What is the final velocity of the bananaplane?

Solution. To find velocity from an acceleration graph, divide the region between the curve (in this case the "curve" is made up of straight lines) and the horizontal axis into triangles and rectangles. See the rectangle and two triangles in the diagram above. Find these areas.

$$A_1 = L_1 W_1 = (20)(-40) = -800 \text{ m}$$
$$A_2 = \frac{1}{2} b_2 h_2 = \frac{1}{2}(6.5)(-40) = -130 \text{ m}$$
$$A_3 = \frac{1}{2} b_3 h_3 = \frac{1}{2}(23.5)(20) = 235 \text{ m}$$

Note that the first two areas are negative because they lie below the t-axis. Add these three areas together to determine the total area.

$$area = A_1 + A_2 + A_3 = -800 - 130 + 235 = -695 \text{ m}$$

Area is **not** the answer. Plug this area into the equation for final velocity:

$$v_x = v_{x0} + area = 150 - 695 = -545 \text{ m/s}$$

The final velocity is approximately $v_x = -545$ m/s. This is the final answer.

(B) When is the acceleration of the bananaplane equal to zero?

Solution. To find acceleration from an acceleration graph, simply read the graph. The acceleration is zero when the "curve" (which in this case is made up of straight lines) crosses the horizontal (t) axis. It first crosses when $t = 26.5$ s. It crosses again at the very end, when $t = 50.0$ s. These are the two answers to this question.

7 TWO OBJECTS IN MOTION

Example 61. Two monkeys, initially 1600 m apart, begin running directly toward one another at the same time. One monkey uniformly accelerates from rest at $\frac{1}{8}$ m/s^2, while the other monkey runs with a constant speed of 15 m/s. What is the net displacement of each monkey when they meet?

Solution. Begin with a labeled diagram. The monkeys begin in different positions, but meet up at the end of the motion.

Identify the knowns, using subscripts where appropriate.

- The first monkey begins from rest, so his initial velocity is zero: $v_{10} = 0$.
- The acceleration of the first monkey is $a_1 = \frac{1}{8}$ m/s^2.
- The velocity of the second monkey is negative because he travels in the opposite direction: $v_{20} = -15$ m/s.
- The second monkey has zero acceleration since his velocity is constant: $a_2 = 0$.
- The initial separation of the monkeys is $d = 1600$ m. (This is **not** net displacement.)

Write an equation for the net displacement of each monkey, using subscripts for any quantities that may be different. Since they spend the same amount of time running, use t for time instead of t_1 and t_2.

$$\Delta x_1 = v_{10}t + \frac{1}{2}a_1t^2$$
$$\Delta x_2 = v_{20}t + \frac{1}{2}a_2t^2$$

Plug the knowns into this equation. To avoid clutter, suppress the units until the end.

$$\Delta x_1 = 0 + \frac{1}{2}\left(\frac{1}{8}\right)t^2 = \frac{1}{16}t^2$$
$$\Delta x_2 = -15t + \frac{1}{2}(0)t^2 = -15t$$

Write an equation of constraint. Together, the two monkeys travel a total distance equal to 1600 m. Since the second monkey's net displacement is negative, we must include a minus sign to make them "add" up to 1600 m (since two minuses make a plus).

$$\Delta x_1 - \Delta x_2 = d$$
$$\Delta x_1 - \Delta x_2 = 1600$$

Substitute the net displacement equations into the equation of constraint.

$$\frac{1}{16}t^2 - (-15t) = 1600$$
$$\frac{1}{16}t^2 + 15t = 1600$$

Recognize that this is a **quadratic equation** because it includes a quadratic term ($\frac{1}{16}t^2$), a linear term ($15t$), and a constant term (1600). Subtract 1600 from both sides such that all three terms are on the same side of the equation, ordered in **standard form**.

$$\frac{1}{16}t^2 + 15t - 1600 = 0$$

Compare this equation to the general form $at^2 + bt + c = 0$ to identify the constants.

$$a = \frac{1}{16} \quad , \quad b = 15 \quad , \quad c = -1600$$

Plug these constants into the **quadratic formula**.

$$t = \frac{-b \pm \sqrt{b^2 - 4ac}}{2a} = \frac{-15 \pm \sqrt{(15)^2 - 4\left(\frac{1}{16}\right)(-1600)}}{2\left(\frac{1}{16}\right)}$$

Note that the two minus signs make a plus sign: $-4\left(\frac{1}{16}\right)(-1600) = +400$.

$$t = \frac{-15 \pm \sqrt{225 + 400}}{\frac{1}{8}} = \frac{-15 \pm \sqrt{625}}{\frac{1}{8}} = \frac{-15 \pm 25}{\frac{1}{8}}$$

We must consider both solutions. Work out the two cases separately.

$$t = \frac{-15 + 25}{\frac{1}{8}} \quad \text{or} \quad t = \frac{-15 - 25}{\frac{1}{8}}$$

$$t = \frac{10}{\frac{1}{8}} \quad \text{or} \quad t = \frac{-40}{\frac{1}{8}}$$

To divide by a fraction, multiply by its **reciprocal**. The reciprocal of $\frac{1}{8}$ equals 8.

$$t = 80 \text{ s} \quad \text{or} \quad t = -320 \text{ s}$$

Since time can't be negative, the correct answer is $t = 80$ s. Plug the time into the equations for net displacement.

$$\Delta x_1 = \frac{1}{16}t^2 = \frac{1}{16}(80)^2 = \frac{1}{16}6400 = 400 \text{ m}$$
$$\Delta x_2 = -15t = -15(80) = -1200 \text{ m}$$

The answers are $\Delta x_1 = 400$ m and $\Delta x_2 = -1200$ m.

Example 62. A monkey steals his uncle's banana and runs away with a constant speed of 9.0 m/s, while his uncle uniformly accelerates from rest at 4.0 m/s². The thief has a 2.0-s headstart. What is the net displacement of each monkey when the thief is caught?

Solution. Begin with a labeled diagram. The monkeys begin and end in the same position.

$$i \bullet \!\!\longrightarrow\!\! f$$
$$+x \longrightarrow$$

Identify the knowns, using subscripts where appropriate.

- The thief's velocity is $v_{10} = 9.0$ m/s.
- The acceleration of the thief is zero because his velocity is constant: $a_1 = 0$.
- The uncle begins from rest, so his initial velocity is zero: $v_{20} = 0$.
- The acceleration of the uncle is $a_2 = 4.0$ m/s².
- The thief's headstart is $\Delta t = 2.0$ s. (This is **not** the time spent running.)

Write an equation for the net displacement of each monkey, using subscripts for any quantities that may be different. Since they have the same net displacement, use Δx for net displacement instead of Δx_1 and Δx_2.

$$\Delta x = v_{10}t_1 + \frac{1}{2}a_1 t_1^2$$

$$\Delta x = v_{20}t_2 + \frac{1}{2}a_2 t_2^2$$

Plug the knowns into this equation. To avoid clutter, suppress the units until the end.

$$\Delta x = 9t_1 + \frac{1}{2}(0)t_1^2 = 9t_1$$

$$\Delta x = 0t_2 + \frac{1}{2}(4)t_2^2 = 2t_2^2$$

Write an equation of constraint. The thief runs for 2.0 s more than his uncle runs, so t_1 (for the thief) is 2.0 s larger than t_2 (for the uncle).

$$t_1 = t_2 + \Delta t$$

Substitute the constraint into the first net displacement equation. Recall that $\Delta t = 2.0$ s.

$$\Delta x = 9(t_2 + \Delta t) = 9t_2 + 9\Delta t = 9t_2 + 9(2) = 9t_2 + 18$$

Set the two net displacement equations equal to each other, since Δx is the same for each. That is, since $\Delta x = 9t_2 + 18$ (from the previous equation) for one monkey and $\Delta x = 2t_2^2$ (find this a few equations back) for the other monkey, it must be true that:

$$9t_2 + 18 = 2t_2^2$$

Recognize that this is a **quadratic equation** because it includes a quadratic term ($2t_2^2$), a linear term ($9t_2$), and a constant term (18). Subtract $2t_2^2$ from both sides such that all three terms are on the same side of the equation, ordered in **standard form**.

$$-2t_2^2 + 9t_2 + 18 = 0$$

Compare this equation to the general form $at^2 + bt + c = 0$ to identify the constants.

$$a = -2 \quad , \quad b = 9 \quad , \quad c = 18$$

Plug these constants into the **quadratic formula**.

$$t_2 = \frac{-b \pm \sqrt{b^2 - 4ac}}{2a} = \frac{-9 \pm \sqrt{(9)^2 - 4(-2)(18)}}{2(-2)}$$

Note that the two minus signs make a plus sign: $-4(-2)(18) = +144$.

$$t_2 = \frac{-9 \pm \sqrt{81 + 144}}{-4} = \frac{-9 \pm \sqrt{225}}{-4} = \frac{-9 \pm 15}{-4}$$

We must consider both solutions. Work out the two cases separately.

$$t_2 = \frac{-9 + 15}{-4} \quad \text{or} \quad t_2 = \frac{-9 - 15}{-4}$$

$$t_2 = \frac{6}{-4} \quad \text{or} \quad t_2 = \frac{-24}{-4}$$

$$t_2 = -1.5 \text{ s} \quad \text{or} \quad t_2 = 6.0 \text{ s}$$

Since time can't be negative, the correct answer is $t_2 = 6.0$ s. Plug the time into the equation for net displacement:

$$\Delta x = 2t_2^2 = 2(6)^2 = 2(36) = 72 \text{ m}$$

Alternatively, we could solve for t_1 to get $t_1 = t_2 + \Delta t = 6 + 2 = 8.0$ s and use the other equation for net displacement to find $\Delta x = 9t_1 = 9(8) = 72$ m. We obtain the same answer either way. The final answer is $\Delta x = 72$ m.

Example 63. A monkey at the top of a 90-m tall cliff parachutes downward with a constant speed of 5.0 m/s at the same time as a monkey at the bottom of the cliff throws a banana straight upward with an initial speed of 40 m/s. Where is the banana when it reaches the parachuting monkey?

Solution. Begin with a labeled diagram. The monkey falls downward while the banana rises upward. As usual for vertical motion problems, we choose $+y$ to point upward.

Identify the knowns, using subscripts where appropriate.

- The monkey's velocity is negative because he falls downward: $v_{10} = -5.0$ m/s.
- The acceleration of the monkey is zero because his velocity is constant: $a_1 = 0$.
- The initial velocity of the banana is $v_{20} = 40$ m/s.
- The banana's acceleration is $a_2 = -9.81$ m/s^2 because it is in free fall.
- The initial separation of the objects is $d = 90$ m. (This is **not** net displacement.)

Write an equation for the net displacement of each object, using subscripts for any quantities that may be different. Since they spend the same amount of time traveling, use t for time instead of t_1 and t_2.

$$\Delta y_1 = v_{10}t + \frac{1}{2}a_1 t^2$$

$$\Delta y_2 = v_{20}t + \frac{1}{2}a_2 t^2$$

Plug the knowns into this equation. To avoid clutter, suppress the units until the end.

$$\Delta y_1 = -5t + \frac{1}{2}(0)t^2 = -5t$$

$$\Delta y_2 = 40t + \frac{1}{2}(-9.81)t^2 = 40t - 4.905t^2$$

Write an equation of constraint. Together, the two objects travel a total distance equal to 90 m. Since the monkey's net displacement is negative, we must include a minus sign to make them "add" up to 90 m (since two minuses make a plus).

$$-\Delta y_1 + \Delta y_2 = d$$

$$-\Delta y_1 + \Delta y_2 = 90$$

Substitute the net displacement equations into the equation of constraint.

$$-(-5t) + 40t - 4.905t^2 = 90$$

Note that the two minus signs make a plus sign: $-(-5t) = +5t$.

$$5t + 40t - 4.905t^2 = 90$$

Combine like terms. The $5t$ and $+40t$ are like terms.

$$45t - 4.905t^2 = 90$$

Recognize that this is a **quadratic equation** because it includes a quadratic term ($-4.905t^2$), a linear term ($45t$), and a constant term (90). Subtract 90 from both sides such that all three terms are on the same side of the equation, ordered in **standard form**.

$$-4.905t^2 + 45t - 90 = 0$$

Compare this equation to the general form $at^2 + bt + c = 0$ to identify the constants.

$$a = -4.905 \quad , \quad b = 45 \quad , \quad c = -90$$

Plug these constants into the **quadratic formula**.

$$t = \frac{-b \pm \sqrt{b^2 - 4ac}}{2a} = \frac{-45 \pm \sqrt{(45)^2 - 4(-4.905)(-90)}}{2(-4.905)}$$

Note that the three minus signs make a minus sign: $-4(-4.905)(-90) = -1765.8$.

$$t = \frac{-45 \pm \sqrt{2025 - 1765.8}}{-9.81} = \frac{-45 \pm \sqrt{259.2}}{-9.81} = \frac{-45 \pm 16.1}{-9.81}$$

We must consider both solutions. Work out the two cases separately.

$$t = \frac{-45 + 16.1}{-9.81} \quad \text{or} \quad t = \frac{-45 - 16.1}{-9.81}$$

$$t = \frac{-28.9}{-9.81} \quad \text{or} \quad t = \frac{-61.1}{-9.81}$$

$$t = 2.95 \text{ s} \quad \text{or} \quad t = 6.23 \text{ s}$$

The first positive time, $t = 2.95$ s, solves the problem. Plug this time into the equations for net displacement.

$$\Delta y_1 = -5t = -5(2.95) = -14.8 \text{ m}$$

$$\Delta y_2 = 40t - 4.905t^2 = 40(2.95) - 4.905(2.95)^2 = 118 - 42.7 = 75.3 \text{ m}$$

The answers are $\Delta y_1 = -15$ m and $\Delta y_2 = 76$ m. (Note that 14.8 m and 75.3 m add up to 90.1 m, which is slightly more than 90 m due to a little rounding.) If you round 9.81 to 10, it's possible to get an approximate answer of 3.0 s for time without using a calculator.

8 NET AND AVERAGE VALUES

Equations	Symbol	Name	SI Units
ave. spd. $= \dfrac{TDT}{TT}$ ave. vel. $= \dfrac{ND}{TT}$ ave. accel. $= \dfrac{v_f - v_i}{TT}$	TT	total time	s
	TDT	total distance traveled	m
	ND	net displacement	m
	ave. spd.	average speed	m/s
	ave. vel.	average velocity	m/s
	ave. accel.	average acceleration	m/s²

Example 64. A monkey travels 15 m east, then 70 m west, and finally 35 m east.

(A) Determine the total distance traveled for the whole trip.

Solution. Begin with a labeled diagram. We choose east for the positive $+x$ direction.

Identify the individual displacements (where east is positive).

- The first displacement is $\Delta x_1 = 15$ m (east).
- The next displacement is $\Delta x_2 = -70$ m (west).
- The last displacement is $\Delta x_3 = 35$ m (east).

Add the given distances in **absolute values** to find the total distance traveled.

$$TDT = |\Delta x_1| + |\Delta x_2| + |\Delta x_3| = |15| + |-70| + |35| = 15 + 70 + 35 = 120 \text{ m}$$

The total distance traveled is $TDT = 120$ m.

(B) Determine the net displacement for the whole trip.

Solution. For net displacement, sign matters: the second displacement is negative.

$$ND = \Delta x_1 + \Delta x_2 + \Delta x_3 = 15 - 70 + 35 = -20 \text{ m}$$

The net displacement is $ND = -20$ m. The minus sign means that the net displacement is 20 m to the west.

Example 65. A gorilla skates 120 m to the west for 3.0 s, and then skates 180 m to the east for 7.0 s.

(A) Find the average speed of the gorilla.

Solution. Begin with a labeled diagram. We choose east for the positive $+x$ direction.

$$i_1 \longleftarrow \qquad \qquad f_1$$
$$f_2 \longrightarrow \qquad \qquad i_2 \qquad +x$$

The definition of average speed is the total distance traveled (TDT) divided by the total time (TT). Therefore, we first need to determine the total distance traveled and the total time for the whole trip. Identify the individual displacements (where east is positive) and the time for each trip.

- The first displacement is $\Delta x_1 = -120$ m (west).
- The last displacement is $\Delta x_2 = 180$ m (east).
- The time for the first displacement is $t_1 = 3.0$ s.
- The time for the last displacement is $t_2 = 7.0$ s.

Add the given distances in **absolute values** to find the total distance traveled.
$$TDT = |\Delta x_1| + |\Delta x_2| = |-120| + |180| = 120 + 180 = 300 \text{ m}$$
Add the given times to find the total time.
$$TT = t_1 + t_2 = 3.0 + 7.0 = 10.0 \text{ s}$$
Now apply the definition of average speed.
$$\frac{\text{ave.}}{\text{spd.}} = \frac{TDT}{TT} = \frac{300}{10} = 30 \text{ m/s}$$
The average speed is 30 m/s. Note that it would be **incorrect** to find the speed for each trip and then find the average of the two different speeds (that method only works if the two times are equal, which isn't the case in this problem).

(B) Find the average velocity of the gorilla.

Solution. The definition of average velocity is the net displacement (ND) divided by the total time (TT). Therefore, we first need to determine the net displacement for the whole trip. For net displacement, sign matters: the first displacement is negative.
$$ND = \Delta x_1 + \Delta x_2 = -120 + 180 = 60 \text{ m}$$
Now apply the definition of average velocity.
$$\frac{\text{ave.}}{\text{vel.}} = \frac{ND}{TT} = \frac{60}{10} = 6.0 \text{ m/s}$$
The average velocity is 6.0 m/s to the east (since the net displacement is to the east).

Example 66. A lemur runs 60 m to the east for 4.0 s, then runs 80 m to the south for 16.0 s.
(A) Find the average speed of the lemur.

Solution. Begin with a labeled diagram. We choose east for the positive $+x$ direction and north for the $+y$ direction.

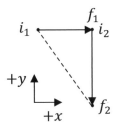

The definition of average speed is the total distance traveled (TDT) divided by the total time (TT). Therefore, we first need to determine the total distance traveled and the total time for the whole trip. Identify the individual displacements (where east and north are positive) and the time for each trip.

- The first displacement is $\Delta x = 60$ m (east).
- The last displacement is $\Delta y = -80$ m (south).
- The time for the first displacement is $t_1 = 4.0$ s.
- The time for the last displacement is $t_2 = 16.0$ s.

Add the given distances in **absolute values** to find the total distance traveled.
$$TDT = |\Delta x| + |\Delta y| = |60| + |-80| = 60 + 80 = 140 \text{ m}$$
Add the given times to find the total time.
$$TT = t_1 + t_2 = 4.0 + 16.0 = 20.0 \text{ s}$$
Now apply the definition of average speed.
$$\text{ave. spd.} = \frac{TDT}{TT} = \frac{140}{20} = 7.0 \text{ m/s}$$
The average speed is 7.0 m/s. Note that it would be **incorrect** to find the speed for each trip and then find the average of the two different speeds (that method only works if the two times are equal, which isn't the case in this problem).

(B) Find the magnitude of the average velocity of the lemur.

Solution. The definition of average velocity is the net displacement (ND) divided by the total time (TT). Therefore, we first need to determine the net displacement for the whole trip. Net displacement is defined as a straight line from the initial position (i_1) to the final position (f_2). It is the hypotenuse of the right triangle illustrated above. Therefore, we must apply the Pythagorean theorem to find the magnitude of the net displacement.
$$ND = \sqrt{\Delta x^2 + \Delta y^2} = \sqrt{(60)^2 + (-80)^2} = \sqrt{3600 + 6400} = \sqrt{10,000} = 100 \text{ m}$$
Now apply the definition of average velocity.
$$\text{ave. vel.} = \frac{ND}{TT} = \frac{100}{20} = 5.0 \text{ m/s}$$
The magnitude of the average velocity is 6.0 m/s. **Note:** The word **magnitude** simply means to ignore the direction of the average velocity, as we will learn in Chapter 10.

Example 67. A gorilla initially traveling 5.0 m/s to the east uniformly accelerates for 4.0 s, by which time the gorilla has a velocity of 25.0 m/s to the east. The gorilla then maintains constant velocity for 10.0 s. Next, the gorilla uniformly decelerates for 6.0 s until coming to rest. Determine the average acceleration of the gorilla.

Solution. Begin with a labeled diagram. We choose east for the positive $+x$ direction.

$$i \bullet\!\!\longrightarrow\!\! f$$
$$+x$$

The definition of average acceleration is final velocity (v_f) minus initial velocity (v_i) all divided by the total time (TT). Therefore, we first need to determine the final velocity, initial velocity, and the total time for the whole trip. Identify the needed information from the problem above.

- The initial velocity is $v_i = 5.0$ m/s (east).
- The final velocity is zero because the gorilla comes to rest $v_f = 0$.
- The time for the first displacement is $t_1 = 4.0$ s.
- The time for the second displacement is $t_2 = 10.0$ s.
- The time for the last displacement is $t_3 = 6.0$ s.

Add the given times to find the total time.

$$TT = t_1 + t_2 + t_3 = 4.0 + 10.0 + 6.0 = 20.0 \text{ s}$$

Now apply the definition of average acceleration.

$$\frac{\text{ave.}}{\text{accel.}} = \frac{v_f - v_i}{TT} = \frac{0 - 5.0}{20} = -\frac{1}{4} \text{ m/s}^2$$

The average acceleration is $-\frac{1}{4}$ m/s^2, which can also be expressed as -0.25 m/s^2. The minus sign means that $\frac{1}{4}$ m/s^2 is to the "west." The average acceleration is opposite to the motion in this case because it is **deceleration**. This is on "average": The gorilla has positive acceleration for trip 1, zero acceleration for trip 2, and negative acceleration (meaning deceleration) for trip 3. Note that it would be **incorrect** (in general) to find the acceleration for each trip and then find the average of the three different accelerations.

Notice that this problem provides extraneous information that does **not** impact the final answer (namely, the intermediate velocity of 25.0 m/s for the middle trip). The reasoning behind this is that you can measure anything you want in the laboratory, but need to be able to figure out which quantities are actually worth measuring (meaning that it's a valuable skill for a few problems to give you a little information that you don't actually need to know in order to solve the problem).

9 REVIEW OF ESSENTIAL TRIGONOMETRY SKILLS

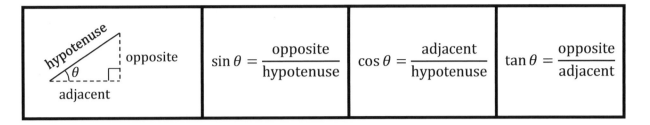

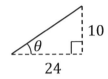	$\sin \theta = \dfrac{\text{opposite}}{\text{hypotenuse}}$	$\cos \theta = \dfrac{\text{adjacent}}{\text{hypotenuse}}$	$\tan \theta = \dfrac{\text{opposite}}{\text{adjacent}}$

Example 68. Find the sine, cosine, and tangent of θ in the diagram below.

Solution. Identify the base and height from the figure.

- The height of the triangle is $a = 10$.
- The base of the triangle is $b = 24$.

We need to find the hypotenuse of the triangle before we can find all of the trig functions. Use the Pythagorean theorem to solve for the hypotenuse.

$$a^2 + b^2 = c^2$$
$$10^2 + 24^2 = c^2$$
$$100 + 576 = 676 = c^2$$

Squareroot both sides to solve for the unknown.

$$c = \sqrt{676} = 26$$

Now identify the sides of the right triangle in relation to the indicated angle θ.

- The side adjacent to θ is adj. $= 24$.
- The side opposite to θ is opp. $= 10$. (This side does not "touch" the angle θ.)
- The hypotenuse is hyp. $= 26$. (The hypotenuse is the longest side.)

Plug these sides into the formulas for the sine, cosine, and tangent functions.

$$\sin \theta = \frac{\text{opp.}}{\text{hyp.}} = \frac{10}{26} = \frac{5}{13} \quad , \quad \cos \theta = \frac{\text{adj.}}{\text{hyp.}} = \frac{24}{26} = \frac{12}{13} \quad , \quad \tan \theta = \frac{\text{opp.}}{\text{adj.}} = \frac{10}{24} = \frac{5}{12}$$

Note that each fraction was reduced by dividing both the numerator and denominator by 2:

$$\frac{10}{26} = \frac{10 \div 2}{26 \div 2} = \frac{5}{13}, \frac{24}{26} = \frac{24 \div 2}{26 \div 2} = \frac{12}{13}, \text{ and } \frac{10}{24} = \frac{10 \div 2}{24 \div 2} = \frac{5}{12}.$$

Example 69. Find the sine, cosine, and tangent of θ in the diagram below.

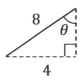

Solution. Identify the base and hypotenuse from the figure.

- The hypotenuse of the triangle is $c = 8$.
- The base of the triangle is $b = 4$.

We need to find the height of the triangle before we can find all of the trig functions. Use the Pythagorean theorem to solve for the height.

$$a^2 + b^2 = c^2$$
$$a^2 + 4^2 = 8^2$$
$$a^2 + 16 = 64$$
$$a^2 = 64 - 16 = 48$$

Squareroot both sides to solve for the unknown. Note that $\sqrt{48}$ isn't in **standard form**. To put $\sqrt{48}$ in standard form, factor out a **perfect square**: 16 is a perfect square since $4^2 = 16$. Apply the rule from algebra that $\sqrt{ab} = \sqrt{a}\sqrt{b}$.

$$a = \sqrt{48} = \sqrt{(16)(3)} = \sqrt{16}\sqrt{3} = 4\sqrt{3}$$

Now identify the sides of the right triangle in relation to the indicated angle θ.

- The side adjacent to θ is adj. $= 4\sqrt{3}$.
- The side opposite to θ is opp. $= 4$. (This side does not "touch" the angle θ.)
- The hypotenuse is hyp. $= 8$. (The hypotenuse is the longest side.)

Plug these sides into the formulas for the sine, cosine, and tangent functions.

$$\sin\theta = \frac{\text{opp.}}{\text{hyp.}} = \frac{4}{8} = \frac{1}{2} \quad , \quad \cos\theta = \frac{\text{adj.}}{\text{hyp.}} = \frac{4\sqrt{3}}{8} = \frac{\sqrt{3}}{2} \quad , \quad \tan\theta = \frac{\text{opp.}}{\text{adj.}} = \frac{4}{4\sqrt{3}} = \frac{1}{\sqrt{3}} = \frac{\sqrt{3}}{3}$$

Note that each fraction was reduced by dividing both the numerator and denominator by the **greatest common factor**: $\frac{4}{8} = \frac{4 \div 4}{8 \div 4} = \frac{1}{2}$, $\frac{4\sqrt{3}}{8} = \frac{4\sqrt{3} \div 4}{8 \div 4} = \frac{\sqrt{3}}{2}$, and $\frac{4}{4\sqrt{3}} = \frac{4 \div 4}{4\sqrt{3} \div 4} = \frac{1}{\sqrt{3}}$. Also note that $\frac{1}{\sqrt{3}}$ isn't in **standard form**. **Rationalize the denominator** in order to express $\frac{1}{\sqrt{3}}$ in standard form. Multiply both the numerator and denominator by $\sqrt{3}$: $\frac{1}{\sqrt{3}} = \frac{1}{\sqrt{3}}\frac{\sqrt{3}}{\sqrt{3}} = \frac{\sqrt{3}}{3}$. Recall that $\sqrt{3}\sqrt{3} = 3$.

Example 70. Determine the sine, cosine, and tangent of $0°$, $30°$, $45°$, $60°$, and $90°$.

Solution. We will use right triangles to figure these out. Let's begin with the $45°$-$45°$-$90°$ right triangle illustrated below.

The two shorter sides (called the legs) are equal because the $45°$ angles are equal. The sides come in the ratio $1:1:\sqrt{2}$ according to the Pythagorean theorem: $1^2 + 1^2 = \left(\sqrt{2}\right)^2$. For this $45°$-$45°$-$90°$ triangle, the opposite and adjacent are both equal to 1, while the hypotenuse equals $\sqrt{2}$. Plug these values into the formulas for the trig functions.

$$\sin 45° = \frac{\text{opp.}}{\text{hyp.}} = \frac{1}{\sqrt{2}} = \frac{\sqrt{2}}{2} \quad , \quad \cos 45° = \frac{\text{adj.}}{\text{hyp.}} = \frac{1}{\sqrt{2}} = \frac{\sqrt{2}}{2} \quad , \quad \tan 45° = \frac{\text{opp.}}{\text{adj.}} = \frac{1}{1} = 1$$

Note that $\frac{1}{\sqrt{2}}$ isn't in **standard form**. **Rationalize the denominator** in order to express $\frac{1}{\sqrt{2}}$ in standard form. Multiply both the numerator and denominator by $\sqrt{2}$: $\frac{1}{\sqrt{2}} = \frac{1}{\sqrt{2}}\frac{\sqrt{2}}{\sqrt{2}} = \frac{\sqrt{2}}{2}$. Recall that $\sqrt{2}\sqrt{2} = 2$.

Now let's look at the $30°$-$60°$-$90°$ right triangle illustrated below.

The ratio of the sides follows from the Pythagorean theorem: $1^2 + \left(\sqrt{3}\right)^2 = 2^2$. For the $30°$ angle, the opposite side is 1, the adjacent side is $\sqrt{3}$, and the hypotenuse is 2.

$$\sin 30° = \frac{\text{opp.}}{\text{hyp.}} = \frac{1}{2} \quad , \quad \cos 30° = \frac{\text{adj.}}{\text{hyp.}} = \frac{\sqrt{3}}{2} \quad , \quad \tan 30° = \frac{\text{opp.}}{\text{adj.}} = \frac{1}{\sqrt{3}} = \frac{\sqrt{3}}{3}$$

Note that $\frac{1}{\sqrt{3}}$ isn't in **standard form**. **Rationalize the denominator** in order to express $\frac{1}{\sqrt{3}}$ in standard form. Multiply both the numerator and denominator by $\sqrt{3}$: $\frac{1}{\sqrt{3}} = \frac{1}{\sqrt{3}}\frac{\sqrt{3}}{\sqrt{3}} = \frac{\sqrt{3}}{3}$. Recall that $\sqrt{3}\sqrt{3} = 3$.

For the $60°$ angle, the opposite side is $\sqrt{3}$, the adjacent side is 1, and the hypotenuse is 2.

$$\sin 60° = \frac{\text{opp.}}{\text{hyp.}} = \frac{\sqrt{3}}{2} \quad , \quad \cos 60° = \frac{\text{adj.}}{\text{hyp.}} = \frac{1}{2} \quad , \quad \tan 60° = \frac{\text{opp.}}{\text{adj.}} = \frac{\sqrt{3}}{1} = \sqrt{3}$$

The trick to evaluating the trig functions at $0°$ is to visualize a small angle θ and apply the limit that θ approaches zero. Consider the right triangle illustrated below, which has a small angle θ and a hypotenuse equal to 1.

As the angle θ approaches zero:

- The opposite side shrinks to 0.
- The adjacent side approaches 1.

$$\sin 0° = \frac{\text{opp.}}{\text{hyp.}} = \frac{0}{1} = 0 \quad , \quad \cos 0° = \frac{\text{adj.}}{\text{hyp.}} = \frac{1}{1} = 1 \quad , \quad \tan 0° = \frac{\text{opp.}}{\text{adj.}} = \frac{0}{1} = 0$$

For $90°$, visualize a large angle θ and apply the limit that θ approaches $90°$. Consider the right triangle illustrated below, which has a large angle θ and a hypotenuse equal to 1.

As the angle θ approaches $90°$:

- The opposite side approaches 1.
- The adjacent side shrinks to 0.

$$\sin 90° = \frac{\text{opp.}}{\text{hyp.}} = \frac{1}{1} = 1 \quad , \quad \cos 90° = \frac{\text{adj.}}{\text{hyp.}} = \frac{0}{1} = 0 \quad , \quad \tan 90° = \frac{\text{opp.}}{\text{adj.}} = \frac{1}{0} = \text{undefined}$$

Note that the tangent of $90°$ is undefined (since the ratio of the opposite to the adjacent approaches infinity as θ approaches $90°$).

All of these trig values are summarized in the following table.

θ	$0°$	$30°$	$45°$	$60°$	$90°$
$\sin \theta$	0	$\frac{1}{2}$	$\frac{\sqrt{2}}{2}$	$\frac{\sqrt{3}}{2}$	1
$\cos \theta$	1	$\frac{\sqrt{3}}{2}$	$\frac{\sqrt{2}}{2}$	$\frac{1}{2}$	0
$\tan \theta$	0	$\frac{\sqrt{3}}{3}$	1	$\sqrt{3}$	undef.

Note: The tangent of $90°$ is undefined.

The Four Quadrants

- If $0° < \theta < 90°$, θ lies in Quadrant I.
- If $90° < \theta < 180°$, θ lies in Quadrant II.
- If $180° < \theta < 270°$, θ lies in Quadrant III.
- If $270° < \theta < 360°$, θ lies in Quadrant IV.

The Signs of the Trig Functions in Each Quadrant

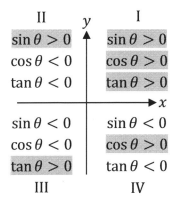

- In Quadrant II, sine is positive, while cosine and tangent are negative.
- In Quadrant III, tangent is positive, while sine and cosine are negative.
- In Quadrant IV, cosine is positive, while sine and tangent are negative.

Finding the Reference Angle

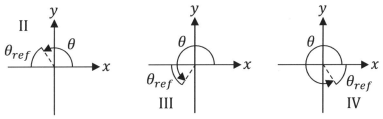

- In Quadrant I, $\theta_{ref} = \theta_I$.
- In Quadrant II, $\theta_{ref} = 180° - \theta_{II}$.
- In Quadrant III, $\theta_{ref} = \theta_{III} - 180°$.
- In Quadrant IV, $\theta_{ref} = 360° - \theta_{IV}$.

Example 71. Evaluate each trig function at the indicated angle.

(A) What is $\sin 150°$?

Solution. The angle $150°$ lies in Quadrant II.

- The sine function is **positive** in Quadrant II.
- Use the Quadrant II formula to find the reference angle.
$$\theta_{ref} = 180° - \theta_{II} = 180° - 150° = 30°$$
- Evaluate sine at the reference angle: $\sin 30° = \frac{1}{2}$.

Combine the sign (+) from step 1 above along with the value from step 3 above.
$$\sin 150° = +\sin 30° = +\frac{1}{2}$$

(B) What is $\cos 240°$?

Solution. The angle $240°$ lies in Quadrant III.

- The cosine function is **negative** in Quadrant III.
- Use the Quadrant III formula to find the reference angle.
$$\theta_{ref} = \theta_{III} - 180° = 240° - 180° = 60°$$
- Evaluate cosine at the reference angle: $\cos 60° = \frac{1}{2}$. This is **not** the final answer.

Combine the sign (−) from step 1 above along with the value from step 3 above.
$$\cos 240° = -\cos 60° = -\frac{1}{2}$$

(C) What is $\tan 300°$?

Solution. The angle $300°$ lies in Quadrant IV.

- The tangent function is **negative** in Quadrant IV.
- Use the Quadrant IV formula to find the reference angle.
$$\theta_{ref} = 360° - \theta_{IV} = 360° - 300° = 60°$$
- Evaluate tangent at the reference angle: $\tan 60° = \sqrt{3}$. This is **not** the final answer.

Combine the sign (−) from step 1 above along with the value from step 3 above.
$$\tan 300° = -\tan 60° = -\sqrt{3}$$

(D) What is $\sin 315°$?

Solution. The angle $315°$ lies in Quadrant IV.

- The sine function is **negative** in Quadrant IV.
- Use the Quadrant IV formula to find the reference angle.
$$\theta_{ref} = 360° - \theta_{IV} = 360° - 315° = 45°$$
- Evaluate sine at the reference angle: $\sin 45° = \frac{\sqrt{2}}{2}$. This is **not** the final answer.

Combine the sign (−) from step 1 above along with the value from step 3 above.
$$\sin 315° = -\sin 45° = -\frac{\sqrt{2}}{2}$$

(E) What is $\cos 135°$?

Solution. The angle $135°$ lies in Quadrant II.

- The cosine function is **negative** in Quadrant II.
- Use the Quadrant II formula to find the reference angle.
$$\theta_{ref} = 180° - \theta_{II} = 180° - 135° = 45°$$
- Evaluate cosine at the reference angle: $\cos 45° = \frac{\sqrt{2}}{2}$. This is **not** the final answer.

Combine the sign $(-)$ from step 1 above along with the value from step 3 above.
$$\cos 135° = -\cos 45° = -\frac{\sqrt{2}}{2}$$

(F) What is $\tan 210°$?

Solution. The angle $210°$ lies in Quadrant III.

- The tangent function is **positive** in Quadrant III.
- Use the Quadrant III formula to find the reference angle.
$$\theta_{ref} = \theta_{III} - 180° = 210° - 180° = 30°$$
- Evaluate tangent at the reference angle: $\tan 30° = \frac{\sqrt{3}}{3}$.

Combine the sign $(+)$ from step 1 above along with the value from step 3 above.
$$\tan 210° = +\tan 30° = +\frac{\sqrt{3}}{3}$$

(G) What is $\sin 240°$?

Solution. The angle $240°$ lies in Quadrant III.

- The sine function is **negative** in Quadrant III.
- Use the Quadrant III formula to find the reference angle.
$$\theta_{ref} = \theta_{III} - 180° = 240° - 180° = 60°$$
- Evaluate sine at the reference angle: $\sin 60° = \frac{\sqrt{3}}{2}$. This is **not** the final answer.

Combine the sign $(-)$ from step 1 above along with the value from step 3 above.
$$\sin 240° = -\sin 60° = -\frac{\sqrt{3}}{2}$$

(H) What is $\cos 180°$?

Solution. The angle $180°$ lies on the border of Quadrants II and III.

- The cosine function is **negative** in both Quadrants II and III.
- Use either the Quadrant II or III formula to find the reference angle. Since $180°$ lies on the border of Quadrants II and III, you get the same answer either way.
$$\theta_{ref} = 180° - \theta_{II} = 180° - 180° = 0°$$
$$\theta_{ref} = \theta_{III} - 180° = 180° - 180° = 0°$$
- Evaluate cosine at the reference angle: $\cos 0° = 1$. This is **not** the final answer.

Combine the sign $(-)$ from step 1 above along with the value from step 3 above.
$$\cos 180° = -\cos 0° = -1$$

(I) What is $\tan 225°$?

Solution. The angle $225°$ lies in Quadrant III.

- The tangent function is **positive** in Quadrant III.
- Use the Quadrant III formula to find the reference angle.
$$\theta_{ref} = \theta_{III} - 180° = 225° - 180° = 45°$$
- Evaluate tangent at the reference angle: $\tan 45° = 1$.

Combine the sign (+) from step 1 above along with the value from step 3 above.
$$\tan 225° = +\tan 45° = +1$$

(J) What is $\sin 180°$?

Solution. The angle $180°$ lies on the border of Quadrants II and III.

- The sine function is positive in Quadrant II, but negative in Quadrant III, which suggests that the final answer will be zero.
- Use either the Quadrant II or III formula to find the reference angle. Since $180°$ lies on the border of Quadrants II and III, you get the same answer either way.
$$\theta_{ref} = 180° - \theta_{II} = 180° - 180° = 0°$$
$$\theta_{ref} = \theta_{III} - 180° = 180° - 180° = 0°$$
- Evaluate sine at the reference angle: $\sin 0° = 0$.

Since the sine of the reference angle equals zero, the sign from step 1 doesn't matter.
$$\sin 180° = \sin 0° = 0$$

(K) What is $\cos 270°$?

Solution. The angle $270°$ lies on the border of Quadrants III and IV.

- The cosine function is negative in Quadrant III, but positive in Quadrant IV, which suggests that the final answer will be zero.
- Use either the Quadrant III or IV formula to find the reference angle. Since $270°$ lies on the border of Quadrants III and IV, you get the same answer either way.
$$\theta_{ref} = \theta_{III} - 180° = 270° - 180° = 90°$$
$$\theta_{ref} = 360° - \theta_{IV} = 360° - 270° = 90°$$
- Evaluate cosine at the reference angle: $\cos 90° = 0$.

Since the cosine of the reference angle equals zero, the sign from step 1 doesn't matter.
$$\cos 270° = \cos 90° = 0$$

(L) What is $\tan 300°$?

Solution. The angle $300°$ lies in Quadrant IV.

- The tangent function is **negative** in Quadrant IV.
- Use the Quadrant IV formula to find the reference angle.
$$\theta_{ref} = 360° - \theta_{IV} = 360° - 300° = 60°$$
- Evaluate tangent at the reference angle: $\tan 60° = \sqrt{3}$. This is **not** the final answer.

Combine the sign (−) from step 1 above along with the value from step 3 above.
$$\tan 300° = -\tan 60° = -\sqrt{3}$$

(M) What is $\sin 300°$?

Solution. The angle $300°$ lies in Quadrant IV.

- The sine function is **negative** in Quadrant IV.
- Use the Quadrant IV formula to find the reference angle.

$$\theta_{ref} = 360° - \theta_{IV} = 360° - 300° = 60°$$

- Evaluate sine at the reference angle: $\sin 60° = \frac{\sqrt{3}}{2}$. This is **not** the final answer.

Combine the sign $(-)$ from step 1 above along with the value from step 3 above.

$$\sin 300° = -\sin 60° = -\frac{\sqrt{3}}{2}$$

(N) What is $\cos 330°$?

Solution. The angle $330°$ lies in Quadrant IV.

- The cosine function is **positive** in Quadrant IV.
- Use the Quadrant IV formula to find the reference angle.

$$\theta_{ref} = 360° - \theta_{IV} = 360° - 330° = 30°$$

- Evaluate cosine at the reference angle: $\cos 30° = \frac{\sqrt{3}}{2}$.

Combine the sign $(+)$ from step 1 above along with the value from step 3 above.

$$\cos 330° = +\cos 30° = +\frac{\sqrt{3}}{2}$$

(O) What is $\tan 180°$?

Solution. The angle $180°$ lies on the border of Quadrants II and III.

- The tangent function is negative in Quadrant II, but positive in Quadrant III, which suggests that the final answer will be zero.
- Use either the Quadrant II or III formula to find the reference angle. Since $180°$ lies on the border of Quadrants II and III, you get the same answer either way.

$$\theta_{ref} = 180° - \theta_{II} = 180° - 180° = 0°$$
$$\theta_{ref} = \theta_{III} - 180° = 180° - 180° = 0°$$

- Evaluate tangent at the reference angle: $\tan 0° = 0$.

Since the tangent of the reference angle equals zero, the sign from step 1 doesn't matter.

$$\tan 180° = \tan 0° = 0$$

(P) What is $\sin 270°$?

Solution. The angle $270°$ lies on the border of Quadrants III and IV.

- The sine function is **negative** in both Quadrants III and IV.
- Use either the Quadrant III or IV formula to find the reference angle. Since $270°$ lies on the border of Quadrants III and IV, you get the same answer either way.

$$\theta_{ref} = \theta_{III} - 180° = 270° - 180° = 90°$$
$$\theta_{ref} = 360° - \theta_{IV} = 360° - 270° = 90°$$

- Evaluate sine at the reference angle: $\sin 90° = 1$. This is **not** the final answer.

Combine the sign $(-)$ from step 1 above along with the value from step 3 above.

$$\sin 270° = -\sin 90° = -1$$

(Q) What is $\cos 225°$?

Solution. The angle $225°$ lies in Quadrant III.

- The cosine function is **negative** in Quadrant III.
- Use the Quadrant III formula to find the reference angle.

$$\theta_{ref} = \theta_{III} - 180° = 225° - 180° = 45°$$

- Evaluate cosine at the reference angle: $\cos 45° = \frac{\sqrt{2}}{2}$. This is **not** the final answer.

Combine the sign $(-)$ from step 1 above along with the value from step 3 above.

$$\cos 225° = -\cos 45° = -\frac{\sqrt{2}}{2}$$

(R) What is $\tan 150°$?

Solution. The angle $150°$ lies in Quadrant II.

- The tangent function is **negative** in Quadrant II.
- Use the Quadrant II formula to find the reference angle.

$$\theta_{ref} = 180° - \theta_{II} = 180° - 150° = 30°$$

- Evaluate tangent at the reference angle: $\tan 30° = \frac{\sqrt{3}}{3}$. This is **not** the final answer.

Combine the sign $(-)$ from step 1 above along with the value from step 3 above.

$$\tan 150° = -\tan 30° = -\frac{\sqrt{3}}{3}$$

Example 72. Evaluate each inverse trig function.

(A) What is $\sin^{-1}\left(-\frac{\sqrt{3}}{2}\right)$?

Solution. First find the reference angle corresponding to $\sin\theta_{ref} = \frac{\sqrt{3}}{2}$.

- The reference angle is $60°$ because $\sin 60° = \frac{\sqrt{3}}{2}$. This is **not** the final answer.

- The argument of $\sin^{-1}\left(-\frac{\sqrt{3}}{2}\right)$ is negative.

- The sine function is **negative** in Quadrants III and IV.

- Use the Quadrant III and IV formulas to solve for θ.
$$\theta_{III} = 180° + \theta_{ref} = 180° + 60° = 240°$$
$$\theta_{IV} = 360° - \theta_{ref} = 360° - 60° = 300°$$

The two answers are $\sin^{-1}\left(-\frac{\sqrt{3}}{2}\right) = 240°$ or $300°$.

(B) What is $\cos^{-1}\left(\frac{\sqrt{2}}{2}\right)$?

Solution. First find the reference angle corresponding to $\cos\theta_{ref} = \frac{\sqrt{2}}{2}$.

- The reference angle is $45°$ because $\cos 45° = \frac{\sqrt{2}}{2}$. This is **not** the final answer.

- The argument of $\cos^{-1}\left(\frac{\sqrt{2}}{2}\right)$ is positive.

- The cosine function is **positive** in Quadrants I and IV.

- Use the Quadrant I and IV formulas to solve for θ.
$$\theta_I = \theta_{ref} = 45°$$
$$\theta_{IV} = 360° - \theta_{ref} = 360° - 45° = 315°$$

The two answers are $\cos^{-1}\left(\frac{\sqrt{2}}{2}\right) = 45°$ or $315°$.

(C) What is $\tan^{-1}(-1)$?

Solution. First find the reference angle corresponding to $\tan\theta_{ref} = 1$.

- The reference angle is $45°$ because $\tan 45° = 1$. This is **not** the final answer.
- The argument of $\tan^{-1}(-1)$ is negative.
- The tangent function is **negative** in Quadrants II and IV.
- Use the Quadrant II and IV formulas to solve for θ.
$$\theta_{II} = 180° - \theta_{ref} = 180° - 45° = 135°$$
$$\theta_{IV} = 360° - \theta_{ref} = 360° - 45° = 315°$$

The two answers are $\tan^{-1}(-1) = 135°$ or $315°$.

(D) What is $\sin^{-1}\left(\frac{1}{2}\right)$?

Solution. First find the reference angle corresponding to $\sin\theta_{ref} = \frac{1}{2}$.

- The reference angle is 30° because $\sin 30° = \frac{1}{2}$. This is **not** the final answer.
- The argument of $\sin^{-1}\left(\frac{1}{2}\right)$ is positive.
- The sine function is **positive** in Quadrants I and II.
- Use the Quadrant I and II formulas to solve for θ.

$$\theta_I = \theta_{ref} = 30°$$
$$\theta_{II} = 180° - \theta_{ref} = 180° - 30° = 150°$$

The two answers are $\sin^{-1}\left(\frac{1}{2}\right) = 30°$ or $150°$.

(E) What is $\cos^{-1}\left(-\frac{\sqrt{3}}{2}\right)$?

Solution. First find the reference angle corresponding to $\cos\theta_{ref} = \frac{\sqrt{3}}{2}$.

- The reference angle is 30° because $\cos 30° = \frac{\sqrt{3}}{2}$. This is **not** the final answer.
- The argument of $\cos^{-1}\left(-\frac{\sqrt{3}}{2}\right)$ is negative.
- The cosine function is **negative** in Quadrants II and III.
- Use the Quadrant II and III formulas to solve for θ.

$$\theta_{II} = 180° - \theta_{ref} = 180° - 30° = 150°$$
$$\theta_{III} = 180° + \theta_{ref} = 180° + 30° = 210°$$

The two answers are $\cos^{-1}\left(-\frac{\sqrt{3}}{2}\right) = 150°$ or $210°$.

(F) What is $\tan^{-1}\left(-\sqrt{3}\right)$?

Solution. First find the reference angle corresponding to $\tan\theta_{ref} = \sqrt{3}$.

- The reference angle is 60° because $\tan 60° = \sqrt{3}$. This is **not** the final answer.
- The argument of $\tan^{-1}\left(-\sqrt{3}\right)$ is negative.
- The tangent function is **negative** in Quadrants II and IV.
- Use the Quadrant II and IV formulas to solve for θ.

$$\theta_{II} = 180° - \theta_{ref} = 180° - 60° = 120°$$
$$\theta_{IV} = 360° - \theta_{ref} = 360° - 60° = 300°$$

The two answers are $\tan^{-1}\left(-\sqrt{3}\right) = 120°$ or $300°$.

(G) What is $\sin^{-1}\left(\frac{\sqrt{2}}{2}\right)$?

Solution. First find the reference angle corresponding to $\sin\theta_{ref} = \frac{\sqrt{2}}{2}$.

- The reference angle is $45°$ because $\sin 45° = \frac{\sqrt{2}}{2}$. This is **not** the final answer.
- The argument of $\sin^{-1}\left(\frac{\sqrt{2}}{2}\right)$ is positive.
- The sine function is **positive** in Quadrants I and II.
- Use the Quadrant I and II formulas to solve for θ.

$$\theta_I = \theta_{ref} = 45°$$
$$\theta_{II} = 180° - \theta_{ref} = 180° - 45° = 135°$$

The two answers are $\sin^{-1}\left(\frac{\sqrt{2}}{2}\right) = 45°$ or $135°$.

(H) What is $\cos^{-1}\left(\frac{1}{2}\right)$?

Solution. First find the reference angle corresponding to $\cos\theta_{ref} = \frac{1}{2}$.

- The reference angle is $60°$ because $\cos 60° = \frac{1}{2}$. This is **not** the final answer.
- The argument of $\cos^{-1}\left(\frac{1}{2}\right)$ is positive.
- The cosine function is **positive** in Quadrants I and IV.
- Use the Quadrant I and IV formulas to solve for θ.

$$\theta_I = \theta_{ref} = 60°$$
$$\theta_{IV} = 360° - \theta_{ref} = 360° - 60° = 300°$$

The two answers are $\cos^{-1}\left(\frac{1}{2}\right) = 60°$ or $300°$.

(I) What is $\tan^{-1}\left(\sqrt{3}\right)$?

Solution. First find the reference angle corresponding to $\tan\theta_{ref} = \sqrt{3}$.

- The reference angle is $60°$ because $\tan 60° = \sqrt{3}$. This is **not** the final answer.
- The argument of $\tan^{-1}\left(\sqrt{3}\right)$ is positive.
- The tangent function is **positive** in Quadrants I and III.
- Use the Quadrant I and III formulas to solve for θ.

$$\theta_I = \theta_{ref} = 60°$$
$$\theta_{III} = 180° + \theta_{ref} = 180° + 60° = 240°$$

The two answers are $\tan^{-1}\left(\sqrt{3}\right) = 60°$ or $240°$.

(J) What is $\sin^{-1}(1)$?

Solution. First find the reference angle corresponding to $\sin \theta_{ref} = 1$.

- The reference angle is $90°$ because $\sin 90° = 1$.
- The argument of $\sin^{-1}(1)$ is positive.
- The sine function is **positive** in Quadrants I and II.
- Use the Quadrant I and II formulas to solve for θ.

$$\theta_I = \theta_{ref} = 90°$$
$$\theta_{II} = 180° - \theta_{ref} = 180° - 90° = 90°$$

Therefore, there is only one answer to this question: $\sin^{-1}(1) = 90°$.

(K) What is $\cos^{-1}\left(\frac{\sqrt{3}}{2}\right)$?

Solution. First find the reference angle corresponding to $\cos \theta_{ref} = \frac{\sqrt{3}}{2}$.

- The reference angle is $30°$ because $\cos 30° = \frac{\sqrt{3}}{2}$. This is **not** the final answer.
- The argument of $\cos^{-1}\left(\frac{\sqrt{3}}{2}\right)$ is positive.
- The cosine function is **positive** in Quadrants I and IV.
- Use the Quadrant I and IV formulas to solve for θ.

$$\theta_I = \theta_{ref} = 30°$$
$$\theta_{IV} = 360° - \theta_{ref} = 360° - 30° = 330°$$

The two answers are $\cos^{-1}\left(\frac{\sqrt{3}}{2}\right) = 30°$ or $330°$.

(L) What is $\tan^{-1}(1)$?

Solution. First find the reference angle corresponding to $\tan \theta_{ref} = 1$.

- The reference angle is $45°$ because $\tan 45° = 1$. This is **not** the final answer.
- The argument of $\tan^{-1}(1)$ is positive.
- The tangent function is **positive** in Quadrants I and III.
- Use the Quadrant I and III formulas to solve for θ.

$$\theta_I = \theta_{ref} = 45°$$
$$\theta_{III} = 180° + \theta_{ref} = 180° + 45° = 225°$$

The two answers are $\tan^{-1}(1) = 45°$ or $225°$.

(M) What is $\sin^{-1}\left(\frac{\sqrt{3}}{2}\right)$?

Solution. First find the reference angle corresponding to $\sin\theta_{ref} = \frac{\sqrt{3}}{2}$.

- The reference angle is $60°$ because $\sin 60° = \frac{\sqrt{3}}{2}$. This is **not** the final answer.
- The argument of $\sin^{-1}\left(\frac{\sqrt{3}}{2}\right)$ is positive.
- The sine function is **positive** in Quadrants I and II.
- Use the Quadrant I and II formulas to solve for θ.
$$\theta_I = \theta_{ref} = 60°$$
$$\theta_{II} = 180° - \theta_{ref} = 180° - 60° = 120°$$

The two answers are $\sin^{-1}\left(\frac{\sqrt{3}}{2}\right) = 60°$ or $120°$.

(N) What is $\cos^{-1}(-1)$?

Solution. First find the reference angle corresponding to $\cos\theta_{ref} = 1$.

- The reference angle is $0°$ because $\cos 0° = 1$. This is **not** the final answer.
- The argument of $\cos^{-1}(-1)$ is negative.
- The cosine function is **negative** in Quadrants II and III.
- Use the Quadrant II and III formulas to solve for θ.
$$\theta_{II} = 180° - \theta_{ref} = 180° - 0° = 180°$$
$$\theta_{III} = 180° + \theta_{ref} = 180° + 0° = 180°$$

Therefore, there is only one answer to this question: $\cos^{-1}(-1) = 180°$.

(O) What is $\tan^{-1}(0)$?

Solution. First find the reference angle corresponding to $\tan\theta_{ref} = 0$.

- The reference angle is $0°$ because $\tan 0° = 0$. This is **not** the final answer.
- The argument of $\tan^{-1}(0)$ is doesn't have a sign. Therefore, the angle could be in any Quadrant.
- Use the Quadrant I, II, III, and IV formulas to solve for θ.
$$\theta_I = \theta_{ref} = 0°$$
$$\theta_{II} = 180° - \theta_{ref} = 180° - 0° = 180°$$
$$\theta_{III} = 180° + \theta_{ref} = 180° + 0° = 180°$$
$$\theta_{IV} = 360° - \theta_{ref} = 360° - 0° = 360°$$

The two answers are $\tan^{-1}(0) = 0°$ or $180°$. Note that $0°$ and $360°$ are the **same** angle.

(P) What is $\sin^{-1}(0)$?

Solution. First find the reference angle corresponding to $\sin\theta_{ref} = 0$.

- The reference angle is $0°$ because $\sin 0° = 0$. This is **not** the final answer.
- The argument of $\sin^{-1}(0)$ is doesn't have a sign. Therefore, the angle could be in any Quadrant.
- Use the Quadrant I, II, III, and IV formulas to solve for θ.

$$\theta_I = \theta_{ref} = 0°$$
$$\theta_{II} = 180° - \theta_{ref} = 180° - 0° = 180°$$
$$\theta_{III} = 180° + \theta_{ref} = 180° + 0° = 180°$$
$$\theta_{IV} = 360° - \theta_{ref} = 360° - 0° = 360°$$

The two answers are $\sin^{-1}(0) = 0°$ or $180°$. Note that $0°$ and $360°$ are the **same** angle.

Secant, Cosecant, and Cotangent

$$\sec\theta = \frac{1}{\cos\theta} \quad , \quad \csc\theta = \frac{1}{\sin\theta} \quad , \quad \cot\theta = \frac{1}{\tan\theta}$$

Derivatives of Trig Functions

$$\frac{d}{d\theta}\sin\theta = \cos\theta \quad , \quad \frac{d}{d\theta}\cos\theta = -\sin\theta \quad , \quad \frac{d}{d\theta}\tan\theta = \sec^2\theta$$

$$\frac{d}{d\theta}\sec\theta = \sec\theta\tan\theta \quad , \quad \frac{d}{d\theta}\csc\theta = -\csc\theta\cot\theta \quad , \quad \frac{d}{d\theta}\cot\theta = -\csc^2\theta$$

Integrals of Trig Functions

$$\int \sin\theta\, d\theta = -\cos\theta + c \quad , \quad \int \cos\theta\, d\theta = \sin\theta + c$$

$$\int \tan\theta\, d\theta = -\ln|\cos\theta| + c \quad , \quad \int \sec\theta\, d\theta = \ln|\sec\theta + \tan\theta| + c$$

$$\int \csc\theta\, d\theta = -\ln|\csc\theta + \cot\theta| + c \quad , \quad \int \cot\theta\, d\theta = \ln|\sin\theta| + c$$

Example 73. Perform the following definite integral.

$$\int_{\theta=180°}^{360°} \sin\theta\, d\theta$$

Solution. First find the anti-derivative of $\sin\theta$:

$$\int_{\theta=180°}^{360°} \sin\theta\, d\theta = [-\cos\theta]_{\theta=180°}^{360°}$$

The notation $[-\cos\theta]_{\theta=180°}^{360°}$ means to evaluate the function $-\cos\theta$ at $\theta = 360°$, then evaluate the function $-\cos\theta$ at $\theta = 180°$, and subtract these results:

$$\int_{\theta=180°}^{360°} \sin\theta\, d\theta = [-\cos\theta]_{\theta=180°}^{360°} = -\cos(360°) - [-\cos(180°)]$$

$$= -\cos(360°) + \cos(180°) = -1 - 1 = -2$$

The answer is $\int_{\theta=180°}^{360°} \sin\theta\, d\theta = -2$.

Example 74. Perform the following definite integral.

$$\int_{\theta=120°}^{240°} \cos\theta \, d\theta$$

Solution. First find the anti-derivative of $\cos\theta$:

$$\int_{\theta=120°}^{240°} \cos\theta \, d\theta = [\sin\theta]_{\theta=120°}^{240°}$$

The notation $[\sin\theta]_{\theta=120°}^{240°}$ means to evaluate the function $\sin\theta$ at $\theta = 240°$, then evaluate the function $\sin\theta$ at $\theta = 120°$, and subtract these results:

$$\int_{\theta=120°}^{240°} \cos\theta \, d\theta = [\sin\theta]_{\theta=120°}^{240°} = \sin(240°) - \sin(120°) = -\frac{\sqrt{3}}{2} - \frac{\sqrt{3}}{2} = -\sqrt{3}$$

The answer is $\int_{\theta=120°}^{240°} \cos\theta \, d\theta = -\sqrt{3}$.

10 VECTOR ADDITION

Components of the Given Vectors	Components of the Resultant Vector	Magnitude and Direction of the Resultant Vector
$A_x = A \cos \theta_A$ $A_y = A \sin \theta_A$ $B_x = B \cos \theta_B$ $B_y = B \sin \theta_B$	$R_x = A_x + B_x$ $R_y = A_y + B_y$	$R = \sqrt{R_x^2 + R_y^2}$ $\theta_R = \tan^{-1}\left(\dfrac{R_y}{R_x}\right)$

Example 75. The monkey vector, $\vec{M}$, has a magnitude of 36 ☺ and direction of 240°. The banana vector, $\vec{B}$, has a magnitude of 18 ☺ and direction of 120°. Find the magnitude and direction of the resultant vector, $\vec{R}$.

Solution. Begin by sketching the vector addition. Draw the given vectors **tip-to-tail**. The vector $\vec{M}$ lies in Quadrant III (since $\theta_M = 240°$), so $\vec{M}$ points down to the left. The vector $\vec{B}$ lies in Quadrant II (since $\theta_B = 120°$), so $\vec{B}$ points up to the left. Also, $\vec{B}$ is half as long as $\vec{M}$, since $B = 18$ ☺ and $M = 36$ ☺. The resultant vector, $\vec{R}$, extends from the tail of $\vec{M}$ to the tip of $\vec{B}$ (based on how we drew $\vec{M}$ and $\vec{B}$ tip-to-tail in the diagram below).

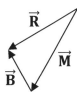

First apply trig to find the components of the given vectors.

$$M_x = M \cos \theta_M = 36 \cos 240° = -36 \cos 60° = -36\left(\frac{1}{2}\right) = -18 \; ☺$$

$$M_y = M \sin \theta_M = 36 \sin 240° = -36 \sin 60° = -36\left(\frac{\sqrt{3}}{2}\right) = -18\sqrt{3} \; ☺$$

$$B_x = B \cos \theta_B = 18 \cos 120° = -18 \cos 60° = -18\left(\frac{1}{2}\right) = -9 \; ☺$$

$$B_y = B \sin \theta_B = 18 \sin 120° = +18 \sin 60° = 18\left(\frac{\sqrt{3}}{2}\right) = 9\sqrt{3} \; ☺$$

Note that 60° is the reference angle for both 120° and 240°. Also, note that cosine is negative in Quadrants II and III, while sine is positive in Quadrant II and negative in Quadrant III. You can also determine the signs from the physics, rather than the trig: Since $\vec{M}$ points down to the left, both M_x and M_y are negative, and since $\vec{B}$ points up to the left, B_x is negative while B_y is positive. Note, for example, that $36\left(\frac{\sqrt{3}}{2}\right) = \frac{36}{2}\sqrt{3} = 18\sqrt{3}$.

Next add the respective components of the given vectors together to find the components of the resultant vector. To calculate $-18\sqrt{3} + 9\sqrt{3}$, **combine like terms** or **factor** out the $\sqrt{3}$: $-18\sqrt{3} + 9\sqrt{3} = (-18 + 9)\sqrt{3} = -9\sqrt{3}$.

$$R_x = M_x + B_x = -18 - 9 = -27 \; ☺$$
$$R_y = M_y + B_y = -18\sqrt{3} + 9\sqrt{3} = -9\sqrt{3} \; ☺$$

To determine the magnitude of the resultant vector, apply the Pythagorean theorem.

$$R = \sqrt{R_x^2 + R_y^2} = \sqrt{(-27)^2 + \left(-9\sqrt{3}\right)^2} = \sqrt{729 + 243} = \sqrt{972} = \sqrt{(18)^2(3)} = 18\sqrt{3} \; ☺$$

Note that the minus signs cancel out since they are squared: $(-1)^2 = +1$. Also note that $\left(\sqrt{3}\right)^2 = 3$ and $\left(9\sqrt{3}\right)^2 = 9^2(3) = 243$. In the last step, we applied the rules from algebra that $\sqrt{ab} = \sqrt{a}\sqrt{b}$ and $\sqrt{x^2} = x$, realizing that 972 can be factored as $972 = 18 \times 18 \times 3 = (18)^2 \times 3$. To determine the direction of the resultant vector, use an inverse tangent.

$$\theta_R = \tan^{-1}\left(\frac{R_y}{R_x}\right) = \tan^{-1}\left(\frac{-9\sqrt{3}}{-27}\right) = \tan^{-1}\left(\frac{\sqrt{3}}{3}\right)$$

The reference angle is $30°$ since $\tan 30° = \frac{\sqrt{3}}{3}$. Tangent is positive in Quadrants I and III. Therefore, the answer lies in Quadrant I or Quadrant III. Which is it? We can deduce that the resultant vector lies in Quadrant III because $R_x < 0$ and $R_y < 0$. Use the Quadrant III formula to find the direction of the resultant vector.

$$\theta_R = 180° + \theta_{ref} = 180° + 30° = 210°$$

The final answers are $R = 18\sqrt{3} \; ☺$ (or $31.2 \; ☺$) and $\theta_R = 210°$.

Example 76. The gorilla vector, $\vec{\Psi}$, has a magnitude of $3\sqrt{2}$ N and direction of $135°$. The chimpanzee vector, $\vec{\Phi}$, has a magnitude of $6\sqrt{2}$ N and direction of $225°$. The orangutan vector, $\vec{\Omega}$, has a magnitude of 12 N and direction of $0°$. Find the magnitude and direction of the resultant vector, $\vec{R}$.

Solution. Begin by sketching the vector addition. Draw all three given vectors **tip-to-tail**. The resultant vector, $\vec{R}$, extends from the tail of $\vec{\Psi}$ to the tip of $\vec{\Omega}$.

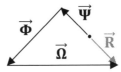

First apply trig to find the components of the given vectors.

$$\Psi_x = \Psi \cos\theta_\Psi = 3\sqrt{2}\cos 135° = -3\sqrt{2}\cos 45° = -3\sqrt{2}\left(\frac{\sqrt{2}}{2}\right) = -3 \text{ N}$$

$$\Psi_y = \Psi \sin\theta_\Psi = 3\sqrt{2}\sin 135° = +3\sqrt{2}\sin 45° = 3\sqrt{2}\left(\frac{\sqrt{2}}{2}\right) = 3 \text{ N}$$

$$\Phi_x = \Phi \cos\theta_\Phi = 6\sqrt{2}\cos 225° = -6\sqrt{2}\cos 45° = -6\sqrt{2}\left(\frac{\sqrt{2}}{2}\right) = -6 \text{ N}$$

$$\Phi_y = \Phi \sin\theta_\Phi = 6\sqrt{2}\sin 225° = -6\sqrt{2}\sin 45° = -6\sqrt{2}\left(\frac{\sqrt{2}}{2}\right) = -6 \text{ N}$$

$$\Omega_x = \Omega \cos\theta_\Omega = 12\cos 0° = 12(1) = 12 \text{ N}$$
$$\Omega_y = \Omega \sin\theta_\Omega = 12\sin 0° = 12(0) = 0 \text{ N}$$

Note, for example, that $3\sqrt{2}\left(\frac{\sqrt{2}}{2}\right) = 3\left(\frac{2}{2}\right) = 3(1) = 3$ because $\sqrt{2}\sqrt{2} = 2$. Next add the respective components of the given vectors together to find the components of the resultant vector.

$$R_x = \Psi_x + \Phi_x + \Omega_x = -3 - 6 + 12 = 3 \text{ N}$$
$$R_y = \Psi_y + \Phi_y + \Omega_y = 3 - 6 + 0 = -3 \text{ N}$$

To determine the magnitude of the resultant vector, apply the Pythagorean theorem.

$$R = \sqrt{R_x^2 + R_y^2} = \sqrt{(3)^2 + (-3)^2} = \sqrt{9+9} = \sqrt{18} = \sqrt{(9)(2)} = \sqrt{9}\sqrt{2} = 3\sqrt{2} \text{ N}$$

To determine the direction of the resultant vector, use an inverse tangent.

$$\theta_R = \tan^{-1}\left(\frac{R_y}{R_x}\right) = \tan^{-1}\left(\frac{-3}{3}\right) = \tan^{-1}(-1)$$

The reference angle is $45°$ since $\tan 45° = 1$. Tangent is negative in Quadrants II and IV. Therefore, the answer lies in Quadrant II or Quadrant IV. Which is it? We can deduce that the resultant vector lies in Quadrant IV because $R_x > 0$ and $R_y < 0$. Use the Quadrant IV formula to find the direction of the resultant vector.

$$\theta_R = 360° - \theta_{ref} = 360° - 45° = 315°$$

The final answers are $R = 3\sqrt{2}$ N (or 4.24 N) and $\theta_R = 315°$.

Example 77. The monkey fur vector, $\vec{F}$, has a magnitude of 16 m and direction of 270°. The monkey tail vector, $\vec{T}$, has a magnitude of $8\sqrt{2}$ m and direction of 225°. The monkey belly vector, $\vec{B}$, is defined according to $\vec{B} = \vec{F} - \vec{T}$. Find the magnitude and direction of $\vec{B}$.

Solution. First apply trig to find the components of the given vectors.

$$F_x = F\cos\theta_F = 16\cos 270° = 16\cos 90° = 16(0) = 0$$
$$F_y = F\sin\theta_F = 16\sin 270° = -16\sin 90° = -16(1) = -16 \text{ m}$$

$$T_x = T\cos\theta_T = 8\sqrt{2}\cos 225° = -8\sqrt{2}\cos 45° = -8\sqrt{2}\left(\frac{\sqrt{2}}{2}\right) = -8 \text{ m}$$

$$T_y = T\sin\theta_T = 8\sqrt{2}\sin 225° = -8\sqrt{2}\sin 45° = -8\sqrt{2}\left(\frac{\sqrt{2}}{2}\right) = -8 \text{ m}$$

Note that $8\sqrt{2}\left(\frac{\sqrt{2}}{2}\right) = 8\left(\frac{2}{2}\right) = 8(1) = 8$ because $\sqrt{2}\sqrt{2} = 2$. Next **<u>subtract</u>** the respective components of the given vectors. We subtract the components because $\vec{B} = \vec{F} - \vec{T}$.

$$B_x = F_x - T_x = 0 - (-8) = 0 + 8 = 8 \text{ m}$$
$$B_y = F_y - T_y = -16 - (-8) = -16 + 8 = -8 \text{ m}$$

Note that subtracting a negative number equates to adding a positive number. For example, $-16 - (-8) = -16 + 8$. To determine the magnitude of $\vec{B}$, apply the Pythagorean theorem.

$$B = \sqrt{B_x^2 + B_y^2} = \sqrt{(8)^2 + (-8)^2} = \sqrt{64 + 64} = \sqrt{128} = \sqrt{(64)(2)} = \sqrt{64}\sqrt{2} = 8\sqrt{2} \text{ m}$$

To determine the direction of $\vec{B}$, use an inverse tangent.

$$\theta_B = \tan^{-1}\left(\frac{B_y}{B_x}\right) = \tan^{-1}\left(\frac{-8}{8}\right) = \tan^{-1}(-1)$$

The reference angle is 45° since $\tan 45° = 1$. Tangent is negative in Quadrants II and IV. Therefore, the answer lies in Quadrant II or Quadrant IV. Which is it? We can deduce that the vector $\vec{B}$ lies in Quadrant IV because $B_x > 0$ and $B_y < 0$. Use the Quadrant IV formula to find the direction of $\vec{B}$.

$$\theta_B = 360° - \theta_{ref} = 360° - 45° = 315°$$

The final answers are $B = 8\sqrt{2}$ m (or 11.3 m) and $\theta_B = 315°$.

Example 78. The math vector, $\vec{\mathbf{M}}$, has a magnitude of 4.0 N and direction of 210°. The science vector, $\vec{\mathbf{S}}$, has a magnitude of 3.0 N and direction of 150°. The physics vector, $\vec{\mathbf{P}}$, is defined according to $\vec{\mathbf{P}} = 3\vec{\mathbf{M}} - 2\vec{\mathbf{S}}$. Find the magnitude and direction of $\vec{\mathbf{P}}$.

Solution. First apply trig to find the components of the given vectors.

$$M_x = M\cos\theta_M = 4\cos 210° = -4\cos 30° = -4\left(\frac{\sqrt{3}}{2}\right) = -2\sqrt{3}\text{ N}$$

$$M_y = M\sin\theta_M = 4\sin 210° = -4\sin 30° = -4\left(\frac{1}{2}\right) = -2\text{ N}$$

$$S_x = S\cos\theta_S = 3\cos 150° = -3\cos 30° = -3\left(\frac{\sqrt{3}}{2}\right) = -\frac{3\sqrt{3}}{2}\text{ N}$$

$$S_y = S\sin\theta_S = 3\sin 150° = 3\sin 30° = 3\left(\frac{1}{2}\right) = \frac{3}{2}\text{ N}$$

Next apply the given equation $\vec{\mathbf{P}} = 3\vec{\mathbf{M}} - 2\vec{\mathbf{S}}$ to the components of the vectors.

$$P_x = 3M_x - 2S_x = 3(-2\sqrt{3}) - 2\left(-\frac{3\sqrt{3}}{2}\right) = -6\sqrt{3} + 3\sqrt{3} = -3\sqrt{3}\text{ N}$$

$$P_y = 3M_y - 2S_y = 3(-2) - 2\left(\frac{3}{2}\right) = -6 - 3 = -9\text{ N}$$

Note that subtracting a negative number equates to adding a positive number. For example, $3(-2\sqrt{3}) - 2\left(-\frac{3\sqrt{3}}{2}\right) = 3(-2\sqrt{3}) + 2\left(\frac{3\sqrt{3}}{2}\right)$. Also, note that $2\left(\frac{3\sqrt{3}}{2}\right) = 3\sqrt{3}$. To determine the magnitude of $\vec{\mathbf{P}}$, apply the Pythagorean theorem.

$$P = \sqrt{P_x^2 + P_y^2} = \sqrt{\left(-3\sqrt{3}\right)^2 + (-9)^2} = \sqrt{27 + 81} = \sqrt{108} = \sqrt{(36)(3)} = \sqrt{36}\sqrt{3} = 6\sqrt{3}\text{ N}$$

Note that $\left(\sqrt{3}\right)^2 = 3$. Thus, $\left(3\sqrt{3}\right)^2 = 3^2(3) = 27$. To determine the direction of $\vec{\mathbf{P}}$, use an inverse tangent.

$$\theta_P = \tan^{-1}\left(\frac{P_y}{P_x}\right) = \tan^{-1}\left(\frac{-9}{-3\sqrt{3}}\right) = \tan^{-1}\left(\frac{3}{\sqrt{3}}\right) = \tan^{-1}(\sqrt{3})$$

Note that $\frac{3}{\sqrt{3}} = \sqrt{3}$ since $\sqrt{3}\sqrt{3} = 3$. The reference angle is 60° since $\tan 60° = \sqrt{3}$. Tangent is positive in Quadrants I and III. Therefore, the answer lies in Quadrant I or Quadrant III. Which is it? We can deduce that the vector $\vec{\mathbf{P}}$ lies in Quadrant III because $P_x < 0$ and $P_y < 0$. Use the Quadrant III formula to find the direction of $\vec{\mathbf{P}}$.

$$\theta_B = 180° + \theta_{ref} = 180° + 60° = 240°$$

The final answers are $P = 6\sqrt{3}\text{ N}$ (or 10.4 N) and $\theta_P = 240°$.

Example 79. Find the magnitude and direction of the following vector:
$$\vec{J} = \sqrt{3}\,\hat{x} - \hat{y}$$
The unit vector $\hat{x}$ points one unit along the x-axis, while the unit vector $\hat{y}$ points one unit along the y-axis. This means that the vector $\vec{J}$ can be expressed as $\vec{J} = J_x\hat{x} + J_y\hat{y}$. Visually, the vector $\vec{J}$ is equivalent to going J_x units along the x-axis and then going J_y units along the y-axis. To solve this problem, compare $\vec{J} = \sqrt{3}\,\hat{x} - \hat{y}$ to $\vec{J} = J_x\hat{x} + J_y\hat{y}$ in order to determine the components of the given vector. The components are $J_x = \sqrt{3}$ and $J_y = -1$. (Note that $-\hat{y}$ is the same thing as $-1\hat{y}$.) To find the magnitude of $\vec{J}$, apply the Pythagorean theorem.
$$J = \sqrt{J_x^2 + J_y^2} = \sqrt{\left(\sqrt{3}\right)^2 + (-1)^2} = \sqrt{3+1} = \sqrt{4} = 2$$
Note that the minus sign is squared: $(-1)^2 = +1$. To determine the direction of $\vec{J}$, use an inverse tangent.
$$\theta_J = \tan^{-1}\left(\frac{J_y}{J_x}\right) = \tan^{-1}\left(\frac{-1}{\sqrt{3}}\right) = \tan^{-1}\left(-\frac{1}{\sqrt{3}}\frac{\sqrt{3}}{\sqrt{3}}\right) = \tan^{-1}\left(-\frac{\sqrt{3}}{3}\right)$$
Note that $\sqrt{3}\sqrt{3} = 3$. The reference angle is $30°$ since $\tan 30° = \frac{\sqrt{3}}{3}$. Tangent is negative in Quadrants II and IV. Therefore, the answer lies in Quadrant II or Quadrant IV. Which is it? We can deduce that the vector $\vec{J}$ lies in Quadrant IV because $J_x > 0$ and $J_y < 0$. Use the Quadrant IV formula to find the direction of $\vec{J}$.
$$\theta_J = 360° - \theta_{ref} = 360° - 30° = 330°$$
The final answers are $J = 2$ and $\theta_J = 330°$.

Example 80. Vector $\vec{K}$ has a magnitude of 4 and a direction of $120°$. Express the vector $\vec{K}$ in terms of Cartesian unit vectors.

Solution. First find the x- and y-components of $\vec{K}$ using trig.
$$K_x = K\cos\theta_K = 4\cos 120° = 4\left(-\frac{1}{2}\right) = -2$$
$$K_y = K\sin\theta_K = 4\sin 120° = 4\left(\frac{\sqrt{3}}{2}\right) = 2\sqrt{3}$$
Use these components to express $\vec{K}$ in terms of unit vectors.
$$\vec{K} = K_x\hat{x} + K_y\hat{y} = -2\,\hat{x} + 2\sqrt{3}\,\hat{y}$$
The answer is $\vec{K} = -2\,\hat{x} + 2\sqrt{3}\,\hat{y}$.

Example 81. Find the magnitude and direction of the resultant of the following two vectors.

$$\vec{V} = 5\sqrt{3}\,\hat{x} - 32\,\hat{y}$$
$$\vec{W} = 3\sqrt{3}\,\hat{x} + 8\,\hat{y}$$

Solution. Add the respective components together to find the resultant vector. That is, **combine like terms**: $5\sqrt{3}\,\hat{x}$ and $3\sqrt{3}\,\hat{x}$ are like terms, and $-32\,\hat{y}$ and $8\,\hat{y}$ are like terms.

$$\vec{R} = \vec{V} + \vec{W} = 5\sqrt{3}\,\hat{x} - 32\,\hat{y} + 3\sqrt{3}\,\hat{x} + 8\,\hat{y} = 5\sqrt{3}\,\hat{x} + 3\sqrt{3}\,\hat{x} - 32\,\hat{y} + 8\,\hat{y}$$
$$\vec{R} = 8\sqrt{3}\,\hat{x} - 24\hat{y}$$

Note that $5\sqrt{3} + 3\sqrt{3} = (5+3)\sqrt{3} = 8\sqrt{3}$. Compare $\vec{R} = 8\sqrt{3}\,\hat{x} - 24\hat{y}$ to $\vec{R} = R_x\hat{x} + R_y\hat{y}$ to determine the components of the resultant vector: $R_x = 8\sqrt{3}$ and $R_y = -24$. To determine the magnitude of the resultant vector, apply the Pythagorean theorem.

$$R = \sqrt{R_x^2 + R_y^2} = \sqrt{\left(8\sqrt{3}\right)^2 + (-24)^2} = \sqrt{192 + 576} = \sqrt{768} = \sqrt{256}\sqrt{3} = 16\sqrt{3}$$

To determine the direction of the resultant vector, use an inverse tangent.

$$\theta_R = \tan^{-1}\left(\frac{R_y}{R_x}\right) = \tan^{-1}\left(\frac{-24}{8\sqrt{3}}\right) = \tan^{-1}\left(\frac{-3}{\sqrt{3}}\right) = \tan^{-1}(-\sqrt{3})$$

The reference angle is $60°$ since $\tan 60° = \sqrt{3}$. Tangent is negative in Quadrants II and IV. Therefore, the answer lies in Quadrant II or Quadrant IV. Which is it? We can deduce that the resultant vector lies in Quadrant IV because $R_x > 0$ and $R_y < 0$. Use the Quadrant IV formula to find the direction of the resultant vector.

$$\theta_R = 360° - \theta_{ref} = 360° - 60° = 300°$$

The final answers are $R = 16\sqrt{3}$ (or 27.7) and $\theta_R = 300°$.

Example 82. Two vectors are defined as follows.

$$\vec{L} = 7\sqrt{2}\,\hat{x} + 3\sqrt{2}\,\hat{y}$$
$$\vec{R} = 2\sqrt{2}\,\hat{x} - \sqrt{2}\,\hat{y}$$

Determine the magnitude and direction of a third vector $\vec{Q}$ where $\vec{Q} = 3\vec{L} - 4\vec{R}$.

Solution. Apply the given equation $\vec{Q} = 3\vec{L} - 4\vec{R}$. Substitute the expressions for $\vec{L}$ and $\vec{R}$ into the equation for $\vec{Q}$.

$$\vec{Q} = 3\vec{L} - 4\vec{R} = 3\left(7\sqrt{2}\,\hat{x} + 3\sqrt{2}\,\hat{y}\right) - 4\left(2\sqrt{2}\,\hat{x} - \sqrt{2}\,\hat{y}\right)$$

Distribute the 3 and the -4. Note that $-4\left(-\sqrt{2}\,\hat{y}\right) = +4\sqrt{2}\,\hat{y}$.

$$\vec{Q} = 3\left(7\sqrt{2}\,\hat{x}\right) + 3\left(3\sqrt{2}\,\hat{y}\right) - 4\left(2\sqrt{2}\,\hat{x}\right) - 4\left(-\sqrt{2}\,\hat{y}\right) = 21\sqrt{2}\,\hat{x} + 9\sqrt{2}\,\hat{y} - 8\sqrt{2}\,\hat{x} + 4\sqrt{2}\,\hat{y}$$

<u>**Combine like terms**</u>: $21\sqrt{2}\,\hat{x}$ and $-8\sqrt{2}\,\hat{x}$ are like terms, and $9\sqrt{2}\,\hat{y}$ and $4\sqrt{2}\,\hat{y}$ are like terms.

$$\vec{Q} = 13\sqrt{2}\,\hat{x} + 13\sqrt{2}\hat{y}$$

Compare $\vec{Q} = 13\sqrt{2}\,\hat{x} + 13\sqrt{2}\hat{y}$ to $\vec{Q} = Q_x\hat{x} + Q_y\hat{y}$ to determine the components of the vector $\vec{Q}$: $Q_x = 13\sqrt{2}$ and $Q_y = 13\sqrt{2}$. To determine the magnitude of the vector $\vec{Q}$, apply the Pythagorean theorem.

$$Q = \sqrt{Q_x^2 + Q_y^2} = \sqrt{\left(13\sqrt{2}\right)^2 + \left(13\sqrt{2}\right)^2} = \sqrt{338 + 338} = \sqrt{676} = 26$$

To determine the direction of $\vec{Q}$, use an inverse tangent.

$$\theta_Q = \tan^{-1}\left(\frac{Q_y}{Q_x}\right) = \tan^{-1}\left(\frac{13\sqrt{2}}{13\sqrt{2}}\right) = \tan^{-1}(1)$$

The reference angle is $45°$ since $\tan 45° = 1$. Tangent is positive in Quadrants I and III. Therefore, the answer lies in Quadrant I or Quadrant III. Which is it? We can deduce that $\vec{Q}$ lies in Quadrant I because $Q_x > 0$ and $Q_y > 0$. Use the Quadrant I formula to find the direction of the resultant vector (the Quadrant I angle is the same as the reference angle).

$$\theta_Q = \theta_{ref} = 45°$$

The final answers are $Q = 26$ and $\theta_Q = 45°$.

11 PROJECTILE MOTION

Equations		
$v_{x0} = v_0 \cos \theta_0$ $v_{y0} = v_0 \sin \theta_0$	$\Delta x = v_{x0}t$ $\Delta y = v_{y0}t + \dfrac{1}{2}a_y t^2$ $v_y = v_{y0} + a_y t$ $v_y^2 = v_{y0}^2 + 2a_y \Delta y$	$v = \sqrt{v_{x0}^2 + v_y^2}$ $\theta = \tan^{-1}\left(\dfrac{v_y}{v_{x0}}\right)$

Symbol	Name	Units
Δx	net horizontal displacement	m
Δy	net vertical displacement	m
v_0	initial speed	m/s
θ_0	direction of initial velocity	$^\circ$
v_{x0}	initial horizontal component of velocity	m/s
v_{y0}	initial vertical component of velocity	m/s
v_y	final vertical component of velocity	m/s
v	final speed	m/s
θ	direction of final velocity	$^\circ$
a_y	acceleration (which is vertical)	m/s^2
t	time	s

Example 83. An ape standing atop a 60-m tall cliff launches a banana with an initial speed of 40 m/s at an angle of 30° above the horizontal. The banana lands on horizontal ground below. How far does the banana travel horizontally? (Neglect the height of the ape.)
Solution. Begin with a labeled diagram.

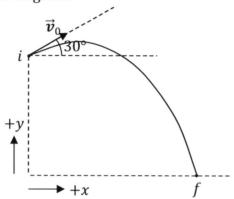

Use the two trig equations to find the components of the initial velocity. Do this **before** listing the known quantities.

$$v_{x0} = v_0 \cos \theta_0 = 40 \cos 30° = 40 \left(\frac{\sqrt{3}}{2} \right) = 20\sqrt{3} \text{ m/s}$$

$$v_{y0} = v_0 \sin \theta_0 = 40 \sin 30° = 40 \left(\frac{1}{2} \right) = 20 \text{ m/s}$$

The unknown we are looking for is Δx. List the four knowns.
- The horizontal component of the initial velocity is $v_{x0} = 20\sqrt{3}$ m/s.
- The vertical component of the initial velocity is $v_{y0} = 20$ m/s.
- The vertical displacement is $\Delta y = -60$ m. It's negative because the final position (f) is **below** the initial position (i).
- The acceleration is $a_y = -9.81$ m/s^2 because the banana is in free fall near earth's surface. It's negative because the banana is accelerating downward.

Since we are looking for Δx, we will need to use the equation $\Delta x = v_{x0}t$. However, we can't use this equation yet because we don't know time. We must therefore find time from one of the other equations first. Given the known quantities listed above, we can use the Δy equation to solve for time.

$$\Delta y = v_{y0}t + \frac{1}{2}a_y t^2$$

Plug the knowns into this equation. To avoid clutter, suppress the units until the end.

$$-60 = 20t + \frac{1}{2}(-9.81)t^2$$

Simplify this equation.

$$-60 = 20t - 4.905t^2$$

Recognize that this is a **quadratic equation** because it includes a quadratic term ($-4.905t^2$), a linear term ($20t$), and a constant term (-60). Use algebra to bring the $-4.905t^2$ and $20t$

to the left-hand side, so that all three terms are on the same side of the equation. (These terms will change sign when we add $4.905t^2$ to both sides and subtract $20t$ from both sides.) Also, order the terms such that the equation is in **standard form**, with the quadratic term first, the linear term second, and the constant term last.

$$4.905t^2 - 20t - 60 = 0$$

Compare this equation to the general form $at^2 + bt + c = 0$ to identify the constants.

$$a = 4.905 \quad , \quad b = -20 \quad , \quad c = -60$$

Plug these constants into the **quadratic formula**.

$$t = \frac{-b \pm \sqrt{b^2 - 4ac}}{2a} = \frac{-(-20) \pm \sqrt{(-20)^2 - 4(4.905)(-60)}}{2(4.905)}$$

Note that $-(-20) = +20$ (two minus signs make a plus sign), $(-20)^2 = +400$ (it's positive since the minus sign is squared), and $-4(4.905)(-60) = +1177.2$ (two minus signs make a plus sign).

$$t = \frac{20 \pm \sqrt{400 + 1177.2}}{9.81} = \frac{20 \pm \sqrt{1577.2}}{9.81} = \frac{20 \pm 39.71}{9.81}$$

We must consider both solutions. Work out the two cases separately.

$$t = \frac{20 + 39.71}{9.81} \quad \text{or} \quad t = \frac{20 - 39.71}{9.81}$$

$$t = \frac{59.71}{9.81} \quad \text{or} \quad t = \frac{-19.71}{9.81}$$

$$t = 6.09 \text{ s} \quad \text{or} \quad t = -2.01 \text{ s}$$

Since time can't be negative, the correct answer is $t = 6.09$ s. (Note that if you round 9.81 to 10, you can get the answers ≈ 6.0 s and ≈ -2.0 s without using a calculator.) Now that we have time, we can use the Δx equation to find the **horizontal** distance traveled.

$$\Delta x = v_{x0}t$$

$$\Delta x = (20\sqrt{3})(6.09)$$

$$\Delta x = 211 \text{ m}$$

The answer is $\Delta x = 211$ m. (If you round a_y to ≈ -10 m/s^2, you can get $\Delta x \approx 120\sqrt{3}$ m without using a calculator.)

Example 84. A monkey throws a banana <u>horizontally</u> off the top of an 80-m tall building with an initial speed of 40 m/s. Neglect the height of the monkey. What is the final velocity of the banana, just before striking the (horizontal) ground below?
Solution. Begin with a labeled diagram.

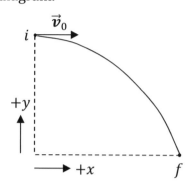

Use the two trig equations to find the components of the initial velocity. Do this **before** listing the known quantities. Since the banana is thrown horizontally, the launch angle is $\theta_0 = 0°$ (because θ_0 is the angle that the initial velocity makes with the horizontal).

$$v_{x0} = v_0 \cos \theta_0 = 40 \cos 0° = 40(1) = 40 \text{ m/s}$$
$$v_{y0} = v_0 \sin \theta_0 = 40 \sin 0° = 40(0) = 0$$

In order to find the final velocity, we first need to find v_y. (Note that v_y is only the vertical component of the final velocity; v_y isn't the final answer.) List the four knowns.

- The horizontal component of the initial velocity is $v_{x0} = 40$ m/s.
- The vertical component of the initial velocity is $v_{y0} = 0$.
- The vertical displacement is $\Delta y = -80$ m. It's negative because the final position (f) is **below** the initial position (i).
- The acceleration is $a_y = -9.81$ m/s^2 because the banana is in free fall near earth's surface. It's negative because the banana is accelerating downward.

Choose the appropriate equation to find v_y from the known quantities listed above.

$$v_y^2 = v_{y0}^2 + 2a_y \Delta y$$
$$v_y^2 = 0^2 + 2(-9.81)(-80)$$

The two minus signs make a plus sign.

$$v_y^2 = 1569.6$$

Squareroot both sides.

$$v_y = \pm 39.6 \text{ m/s}$$

Note that there are two solutions to consider because $(39.6)^2 = 1569.6$ and $(-39.6)^2 = 1569.6$. The correct answer is the **negative** root because the banana is heading **downward** in the final position.

$$v_y = -39.6 \text{ m/s}$$

Note that v_y is **not** the final answer: v_y is just the vertical component of the final velocity. The question asked us to find the final velocity, not just its vertical component. Velocity has two parts: speed and direction. Thus, we need to find the final speed of the banana and the direction of the final velocity. We can apply the Pythagorean theorem to determine the final speed of the banana.

$$v = \sqrt{v_{x0}^2 + v_y^2}$$

It's really $v = \sqrt{v_x^2 + v_y^2}$, but since $v_x = v_{x0}$ for a projectile (since the horizontal component of velocity doesn't change), $\sqrt{v_x^2 + v_y^2}$ and $\sqrt{v_{x0}^2 + v_y^2}$ are the same. However, it's very important to use v_y instead of v_{y0} (since v_{y0} and v_y are very different). Plug numbers into the equation for the final speed.

$$v = \sqrt{(40)^2 + (-39.6)^2} = \sqrt{1600 + 1569.6} = \sqrt{3169.6} = 56.3 \text{ m/s}$$

To determine the direction of the final velocity, use an inverse tangent.

$$\theta_v = \tan^{-1}\left(\frac{v_y}{v_{x0}}\right) = \tan^{-1}\left(\frac{-39.6}{40}\right) = \tan^{-1}(-0.99)$$

The reference angle is 44.7° since $\tan 44.7° = 0.99$. Tangent is negative in Quadrants II and IV. Therefore, the answer lies in Quadrant II or Quadrant IV. Which is it? We can deduce that $\vec{v}$ lies in Quadrant IV because $v_x > 0$ (recall that $v_x = v_{x0}$ for a projectile) and $v_y < 0$. Use the Quadrant IV formula to find the direction of the final velocity.

$$\theta_v = 360° - \theta_{ref} = 360° - 44.7° = 315°$$

The final answers are $v = 56.3$ m/s and $\theta_v = 315°$. (If you round a_y to ≈ -10 m/s^2, you can get $v \approx 40\sqrt{2}$ m/s and $\theta_v = 315°$ without using a calculator.)

Example 85. A monkey throws your physics textbook with an initial speed of 20 m/s at an angle of 30° <u>below</u> the horizontal from the roof of a 75-m tall building. Neglect the monkey's height. How far does the textbook travel horizontally before striking the (horizontal) ground below?

Solution. Begin with a labeled diagram.

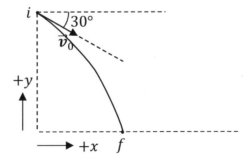

Use the two trig equations to find the components of the initial velocity. Do this **before** listing the known quantities. Note that the launch angle (θ_0) is in **Quadrant IV** because the textbook is thrown **below** the horizontal.

$$v_{x0} = v_0 \cos \theta_0 = 20 \cos 330° = +20 \cos 30° = 20\left(\frac{\sqrt{3}}{2}\right) = 10\sqrt{3} \text{ m/s}$$

$$v_{y0} = v_0 \sin \theta_0 = 20 \sin 330° = -20 \sin 30° = -20\left(\frac{1}{2}\right) = -10 \text{ m/s}$$

The unknown we are looking for is Δx. List the four knowns.

- The horizontal component of the initial velocity is $v_{x0} = 10\sqrt{3}$ m/s.
- The vertical component of the initial velocity is $v_{y0} = -10$ m/s. It's negative because the textbook is heading downward in the initial position.
- The vertical displacement is $\Delta y = -75$ m. It's negative because the final position (f) is **below** the initial position (i).
- The acceleration is $a_y = -9.81$ m/s^2 because the textbook is in free fall near the earth's surface. It's negative because the textbook is accelerating downward.

We can use the Δy equation to solve for time, and then use time to find Δx.

$$\Delta y = v_{y0}t + \frac{1}{2}a_y t^2$$

$$-75 = -10t + \frac{1}{2}(-9.81)t^2$$

$$-75 = -10t - 4.905t^2$$

Recognize that this is a **quadratic equation**. Use algebra to put the equation in **standard form**, with the quadratic term first, the linear term second, and the constant term last.

$$4.905t^2 + 10t - 75 = 0$$

Compare this equation to the general form $at^2 + bt + c = 0$ to identify the constants.

$$a = 4.905 \quad , \quad b = 10 \quad , \quad c = -75$$

Plug these constants into the **quadratic formula**.

$$t = \frac{-b \pm \sqrt{b^2 - 4ac}}{2a} = \frac{-10 \pm \sqrt{(10)^2 - 4(4.905)(-75)}}{2(4.905)}$$

$$t = \frac{-10 \pm \sqrt{100 + 1471.5}}{9.81} = \frac{-10 \pm \sqrt{1571.5}}{9.81} = \frac{-10 \pm 39.64}{9.81}$$

$$t = \frac{-10 + 39.64}{9.81} \quad \text{or} \quad t = \frac{-10 - 39.64}{9.81}$$

$$t = \frac{29.64}{9.81} \quad \text{or} \quad t = \frac{-49.64}{9.81}$$

$$t = 3.02 \text{ s} \quad \text{or} \quad t = -5.06 \text{ s}$$

Since time can't be negative, the correct answer is $t = 3.02$ s. (Note that if you round 9.81 to 10, you can get the answers ≈ 3.0 s and ≈ -5.0 s without using a calculator.) Now that we have time, we can use the Δx equation to find the **horizontal** distance traveled.

$$\Delta x = v_{x0}t$$

$$\Delta x = \left(10\sqrt{3}\right)(3.02)$$

$$\Delta x = 52.3 \text{ m}$$

The answer is $\Delta x = 52.3$ m. (If you round a_y to ≈ -10 m/s^2, then $\Delta x \approx 30\sqrt{3}$ m.)

12 TWO-DIMENSIONAL MOTION WITH CALCULUS

Equations			
$\Delta\vec{\mathbf{r}} = \displaystyle\int_{t=t_0}^{t} \vec{\boldsymbol{v}}\,dt$	$\vec{\boldsymbol{v}} = \vec{\boldsymbol{v}}_0 + \displaystyle\int_{t=t_0}^{t} \vec{\mathbf{a}}\,dt$	$\vec{\boldsymbol{v}} = \dfrac{d\vec{\mathbf{r}}}{dt}$	$\vec{\mathbf{a}} = \dfrac{d\vec{\boldsymbol{v}}}{dt} = \dfrac{d^2\vec{\mathbf{r}}}{dt^2}$

Symbol	Name	SI Units
$\hat{\mathbf{x}}$	unit vector along $+x$	unitless
$\hat{\mathbf{y}}$	unit vector along $+y$	unitless
$\vec{\mathbf{r}}_0 = x_0\,\hat{\mathbf{x}} + y_0\,\hat{\mathbf{y}}$	initial position vector	m
$\vec{\mathbf{r}} = x\,\hat{\mathbf{x}} + y\,\hat{\mathbf{y}}$	position vector	m
$\vec{\mathbf{r}} = \vec{\mathbf{r}} - \vec{\mathbf{r}}_0$ $= (x - x_0)\hat{\mathbf{x}} + (y - y_0)\hat{\mathbf{y}}$	net displacement	m
$\vec{\boldsymbol{v}}_0 = v_{x0}\,\hat{\mathbf{x}} + v_{y0}\,\hat{\mathbf{y}}$	initial velocity vector	m/s
$\vec{\boldsymbol{v}} = v_x\,\hat{\mathbf{x}} + v_y\,\hat{\mathbf{y}}$	velocity vector	m/s
$\vec{\mathbf{a}} = a_x\,\hat{\mathbf{x}} + a_y\,\hat{\mathbf{y}}$	acceleration vector	m/s^2

Example 86. The position of a bananamobile is given according to the following equation, where SI units have been suppressed.

$$\vec{\mathbf{r}} = 2t^4\,\hat{\mathbf{x}} - 4t^3\,\hat{\mathbf{y}}$$

(A) Determine the velocity of the bananamobile at $t = 3.0$ s.

Solution. The given vector ($\vec{\mathbf{r}}$) is called the position vector: It combines the horizontal position (x) and vertical position (y) into a vector of the form $\vec{\mathbf{r}} = x\,\hat{\mathbf{x}} + y\,\hat{\mathbf{y}}$. Compare the given vector $\vec{\mathbf{r}} = 2t^4\,\hat{\mathbf{x}} - 4t^3\,\hat{\mathbf{y}}$ to the general form $\vec{\mathbf{r}} = x\,\hat{\mathbf{x}} + y\,\hat{\mathbf{y}}$ to see that $x = 2t^4$ and $y = -4t^3$. Velocity is the first **derivative** of position with respect to time.

$$\vec{\boldsymbol{v}} = \frac{d\vec{\mathbf{r}}}{dt} = \frac{d}{dt}(2t^4\,\hat{\mathbf{x}} - 4t^3\,\hat{\mathbf{y}})$$

The unit vectors ($\hat{\mathbf{x}}$ and $\hat{\mathbf{y}}$) are constants. In the calculus, treat $\hat{\mathbf{x}}$ and $\hat{\mathbf{y}}$ the same way that you would treat any other constants: You can factor them out of the derivative.

$$\vec{\boldsymbol{v}} = \hat{\mathbf{x}}\frac{d}{dt}(2t^4) - \hat{\mathbf{y}}\frac{d}{dt}(4t^3) = \hat{\mathbf{x}}(4)(2)t^{4-1} - \hat{\mathbf{y}}(3)(4)t^{3-1} = 8t^3\,\hat{\mathbf{x}} - 12t^2\,\hat{\mathbf{y}}$$

Evaluate the velocity function at the specified time ($t = 3.0$ s):
$$\vec{v}(\text{at } t = 3 \text{ s}) = 8(3)^3\,\hat{x} - 12(3)^2\,\hat{y} = 216\,\hat{x} - 108\,\hat{y}$$
The velocity at the specified time is $\vec{v} = 216\,\hat{x} - 108\,\hat{y}$.

(B) Determine the acceleration of the bananamobile at $t = 5.0$ s.

Solution. Acceleration is the first **derivative** of velocity with respect to time. Be careful to take a derivative of the velocity **function** ($\vec{v} = 8t^3\,\hat{x} - 12t^2\,\hat{y}$) that we found in part (A), **not** the value of $\vec{v} = 216\,\hat{x} - 108\,\hat{y}$ found after plugging in time.

$$\vec{a} = \frac{d\vec{v}}{dt} = \frac{d}{dt}(8t^3\,\hat{x} - 12t^2\,\hat{y}) = \hat{x}\frac{d}{dt}(8t^3) - \hat{y}\frac{d}{dt}(12t^2)$$
$$\vec{a} = \hat{x}(3)(8)t^{3-1} - \hat{y}(2)(12)t^{2-1} = 24t^2\,\hat{x} - 24t\,\hat{y}$$

Evaluate the acceleration function at the specified time ($t = 5.0$ s). Note that the time specified in part (B) is **not** the same as the time that was specified in part (A).
$$\vec{a}(\text{at } t = 5 \text{ s}) = 24(5)^2\,\hat{x} - 24(5)\,\hat{y} = 600\,\hat{x} - 120\,\hat{y}$$
The acceleration at the specified time is $\vec{a} = 600\,\hat{x} - 120\,\hat{y}$.

Example 87. The velocity of a bananaplane is given according to the following equation, where SI units have been suppressed.
$$\vec{v} = 8t^3\,\hat{x} - 9t^2\,\hat{y}$$
(A) Determine the net displacement of the bananaplane from $t = 0$ to $t = 3.0$ s.

Solution. Net displacement is the definite **integral** of velocity over time. Treat the unit vectors ($\hat{x}$ and $\hat{y}$) as constants.

$$\Delta\vec{r} = \int_{t=t_0}^{t} \vec{v}\,dt = \int_{t=0}^{3} (8t^3\,\hat{x} - 9t^2\,\hat{y})\,dt = \int_{t=0}^{3} 8t^3\,\hat{x}\,dt - \int_{t=0}^{3} 9t^2\,\hat{y}\,dt$$

$$\Delta\vec{r} = \left[\frac{8t^{3+1}}{3+1}\right]_{t=0}^{3}\hat{x} - \left[\frac{9t^{2+1}}{2+1}\right]_{t=0}^{3}\hat{y} = \left[\frac{8t^4}{4}\right]_{t=0}^{3}\hat{x} - \left[\frac{9t^3}{3}\right]_{t=0}^{3}\hat{y} = [2t^4]_{t=0}^{3}\hat{x} - [3t^3]_{t=0}^{3}\hat{y}$$

$$\Delta\vec{r} = [2(3)^4 - 2(0)^4]\hat{x} - [3(3)^3 - 3(0)^3]\hat{y} = 162\,\hat{x} - 81\,\hat{y}$$

The net displacement for the specified time interval is $\Delta\vec{r} = 162\,\hat{x} - 81\,\hat{y}$.

(B) Determine the acceleration of the bananaplane at $t = 6.0$ s.

Solution. Acceleration is the first **derivative** of velocity with respect to time.
$$\vec{a} = \frac{d\vec{v}}{dt} = \frac{d}{dt}(8t^3\,\hat{x} - 9t^2\,\hat{y}) = \hat{x}\frac{d}{dt}(8t^3) - \hat{y}\frac{d}{dt}(9t^2)$$
$$\vec{a} = \hat{x}(3)(8)t^{3-1} - \hat{y}(2)(9)t^{2-1} = 24t^2\,\hat{x} - 18t\,\hat{y}$$

Evaluate the acceleration function at the specified time ($t = 6.0$ s). Note that the time specified in part (B) is **not** the same as the time that was specified in part (A).
$$\vec{a}(\text{at } t = 6 \text{ s}) = 24(6)^2\,\hat{x} - 18(6)\,\hat{y} = 864\,\hat{x} - 108\,\hat{y}$$
The acceleration at the specified time is $\vec{a} = 864\,\hat{x} - 108\,\hat{y}$.

13 NEWTON'S LAWS OF MOTION

Newton's Laws of Motion

1. According to Newton's first law of motion, every object has a natural tendency (called **inertia**) to maintain **constant momentum** (where momentum equals mass times velocity).

2. According to Newton's second law of motion, the **net force** exerted on an object of constant mass equals the object's **mass times its acceleration**:

$$\sum \vec{\mathbf{F}} = m\vec{\mathbf{a}}$$

For an object with variable mass, like a rocket (Chapter 25), Newton's second law states that the net force equals a derivative of momentum ($\vec{\mathbf{p}}$) with respect to time.

$$\sum \vec{\mathbf{F}} = \frac{d\vec{\mathbf{p}}}{dt}$$

Notation: The uppercase Greek sigma (Σ) is a summation symbol. It states that the left-hand side is the sum of multiple forces (involving vector addition).

3. According to Newton's third law of motion, if one object (call it object A) exerts a force on a second object (call it object B), then object B exerts a force on object A that is **equal** in magnitude and **opposite** in direction to the force that object A exerts on object B. Newton's third law can be expressed concisely in an equation:

$$\vec{\mathbf{F}}_{AB} = -\vec{\mathbf{F}}_{BA}$$

Example 88. Fill in each blank.

(A) An object is accelerating if it is _____.

Solution. Apply the definition of acceleration: Since **acceleration** is the instantaneous rate at which **velocity changes**, the answer is "changing velocity."

(B) A banana will land _____ if a passenger riding inside of a train traveling horizontally to the east with constant velocity drops the banana directly above an X marked on the floor of the train.

Solution. Apply Newton's first law, which involves the concept of **inertia**. The banana was initially moving horizontally. The banana retains this horizontal component to its velocity because the banana has inertia. Throughout the banana's motion, the banana's horizontal velocity matches the train's velocity. When the banana falls, the banana and train travel the same horizontal distance. The answer is "on the X."

(C) A banana will land _____ if a passenger riding inside of a train traveling horizontally to the east while decelerating drops the banana directly above an X marked on the floor of the train.

Solution. Apply Newton's first law, which involves the concept of **inertia**. The banana was initially moving horizontally. The banana retains this horizontal component to its velocity because the banana has inertia. Whereas the banana's horizontal component of velocity is constant, the train's velocity is decreasing because the train is decelerating. The banana's horizontal velocity exceeds the velocity of the train. (The train is slowing down. The banana is not.) The answer is "east of the X."

(D) A football player exerts _____ force on a cheerleader compared to the force that the cheerleader exerts on the football player when a 150-kg football player running 8 m/s collides with a 60-kg cheerleader who was at rest prior to the collision.

Solution. According to Newton's third law, the forces are **equal** in magnitude. The answer is "the same." ("Equal and opposite" would also be okay.)

(E) A football player experiences _____ acceleration during the collision compared to a cheerleader when a 150-kg football player running 8 m/s collides with a 60-kg cheerleader who was at rest prior to the collision.

Solution. We already reasoned that the forces are equal in part (D). The net force on either person is mass times acceleration (Newton's second law). Since the forces are equal, more mass implies less acceleration. The football player has more mass, so he accelerates less during the collision. The answer is "less."

(F) All objects have a natural tendency to maintain _____ acceleration.

Solution. Apply Newton's first law, which involves the concept of **inertia**. Objects have inertia, which is a natural tendency to maintain constant velocity. **Acceleration** is the instantaneous rate at which **velocity changes**. If velocity is constant, acceleration is zero. Therefore, objects have a natural tendency to have **zero** acceleration. The answer is "zero."

(G) A shooter experiences recoil due to Newton's _____ law of motion when firing a rifle. (Definition: A recoil is a push backwards.)

Solution. Apply Newton's third law. When the gun exerts a force on the bullet, the bullet exerts a force back on the gun that is equal in magnitude and **opposite** in direction. The shooter experiences this opposite force (push backwards) due to Newton's third law. The answer is "third."

(H) A cannonball will land _____ when a monkey sailing with a constant velocity of 25 m/s to the south fires a cannonball straight upward relative to the ship.

Solution. Apply Newton's first law, which involves the concept of **inertia**. This question involves the same reasoning as part (B). Compare these questions. The cannonball was initially moving horizontally with the same velocity as the ship. The cannonball retains this horizontal component to its velocity because it has inertia. The cannonball's horizontal velocity matches the ship's velocity. The cannonball and ship travel the same horizontal distance during the cannonball's motion. The answer is "in the cannon."

(I) A 5-g mosquito exerts a force of _____ on a 500-g flyswatter when a monkey swats the mosquito with a force of 25 N.

Solution. According to Newton's third law, the forces are **equal** in magnitude. The answer is "25 N."

Example 89. Fill in each blank.

(A) A 24-kg monkey on earth has a _____ mass and _____ weight on the moon, where gravity is reduced by a factor of six compared to the earth.

Solution. When the monkey visits the moon, the monkey's **mass** remains the **same**, whereas the monkey's weight is reduced by a factor of 6. The given quantity, 24 kg, is clearly the monkey's mass (**not** weight) because the SI unit of mass is the kilogram (kg), whereas the SI unit of weight is the Newton (N). Based on the units, the mass of the monkey is 24 kg. The monkey has the same mass on the moon. Before finding the monkey's weight on the moon, use the equation $W_e = mg_e$ to find the monkey's weight on the earth. The monkey's weight on earth is $W_e = (24)(9.81) = 235$ N. The monkey weighs approximately 6 times less on the moon. Divide W_e by 6 to find the monkey's weight on the moon. The monkey's eight on the moon is $W_m \approx \frac{W_e}{6} = \frac{235}{6} = 39$ N. The answers are "$m = 24$ kg" (for **mass**) and "$W_m = 39$ N" (for **weight** on the moon).

(B) Objects tend to resist changing their _____.

Solution. Apply Newton's first law, which involves the concept of **inertia**. Objects have inertia, which is a natural tendency to maintain constant **momentum**. This means that an object resists changes to its momentum. The answer is "momentum." ("Velocity" is also a good answer. Recall that momentum equals mass times velocity.)

(C) A 600-kg gorilla experiences _____ force and _____ acceleration as a 200-kg monkey when the gorilla collides head-on with the monkey.

Solution. According to Newton's third law, the forces are **equal** in magnitude. The net force on either animal is mass times acceleration (Newton's second law). Since the forces are equal, more mass implies less acceleration. The gorilla has more mass, so the gorilla accelerates less during the collision. The answers are "the same" (for **force**) and "less" (for the gorilla's **acceleration**)

(D) You get _____ when you multiply mass times velocity.

Solution. Consider the equation $\vec{p} = m\vec{v}$, where $\vec{p}$ represents **momentum**. Momentum equals mass times velocity. The answer is "momentum."

(E) You get _____ when you multiply mass times acceleration.

Solution. Consider the equation for Newton's second law: $\sum \vec{F} = m\vec{a}$, where $\sum \vec{F}$ represents net force. **Net force** equals mass times acceleration. The answer is "net force."

(F) The force that the earth exerts on the moon is _____ the force that the moon exerts on the earth.

Solution. According to Newton's third law, the forces are **equal** in magnitude. The answer is "equal to." (It's more precise to say "equal and opposite to.")

(G) A monkey can _____ in order to have nonzero acceleration while running with constant speed.

Solution. Recall that **acceleration** is the instantaneous rate at which velocity changes. Also recall that **velocity** is a combination of speed and direction. Since the monkey's speed is constant, the monkey must change the **direction** of his velocity in order to accelerate. The answer is "change direction." (It would be okay to say "run in a circle.")

(H) A monkey will have zero _____ and constant _____ if the net force acting on the monkey is zero.

Solution. According to Newton's second law, net force equals mass times acceleration. If the net force is zero, the acceleration must be zero also. Since **acceleration** is the instantaneous rate at which velocity changes, zero acceleration implies constant velocity (that is, velocity doesn't change). The answers are "acceleration" and "velocity."

(I) A banana reaches the ground _____ as a feather when the banana and feather are released from the same height above horizontal ground at the same time in a perfect vacuum.

Solution. Look at the equations of free fall (Chapter 3). Observe that **mass** is not a factor in those equations. In a perfect vacuum, all objects fall with the same acceleration, **regardless of mass**. The banana and feather both have an acceleration of $a_y = -9.81 \text{ m/s}^2$. Since they have the same acceleration and descend the same height, and since they are released simultaneously, they reach the ground at the same time. The answer is "at the same time." (Try searching for a feather-penny demo on YouTube.)

(J) A monkey can get home by _____ when the monkey is stranded on horizontal frictionless ice with nothing but a banana.

Solution. Apply Newton's third law to the banana. If the monkey exerts a force on the banana by throwing it, what will happen? According to Newton's third law, the banana exerts a force back on the monkey that is equal in magnitude and opposite in direction to the force that the monkey exerts on the banana. The monkey should throw the banana directly away from his house so that the equal and opposite reaction will push the monkey toward his house. The answer is "throwing the banana directly away from his house."

Example 90. Select the best answer to each question.

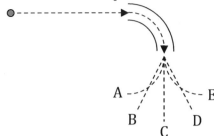

(A) The diagram above shows the top view of a miniature golf hole. A golf ball rolls along a circular metal arc. The dashed line represents the path of the golf ball. When the golf ball loses contact with the metal arc, which path will it follow?

Solution. Apply Newton's first law, which involves the concept of **inertia**. Objects have **inertia**, which is a natural tendency to travel with constant velocity. This means to travel with constant speed in a **straight line**. After leaving the metal arc, there is no longer a force pushing on the golf ball that will change the direction of its velocity. After leaving the metal arc, the golf ball's path is determined by its inertia. The golf ball will continue in a straight line tangent to the metal arc (path C). The answer is C.

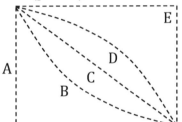

(B) The diagram above illustrates an airplane flying horizontally to the right. The airplane releases a box of bananas from rest (relative to the airplane) when it is at the top left position of the picture above. Which path will the box of bananas follow?

Solution. Apply Newton's first law, which involves the concept of **inertia**. The box of bananas has **inertia** – a natural tendency to maintain constant velocity. Initially, the box of bananas has a horizontal velocity equal to the plane's velocity. When the box of bananas is released, it acquires a vertical component of velocity to go along with its initial horizontal velocity. Since we neglect air resistance unless otherwise stated, the horizontal component of the box's velocity will remain constant due to the box's inertia. The box's vertical component of velocity will increase due to gravity. Which path begins horizontally (due to the box's inertia), and tilts more and more vertically as time goes on? That is path D. Note that the answer would be the same if a gorilla stood at the top of a tall building and threw the box horizontally. The box is a **projectile**. It will follow the path of projectile motion (Chapter 11), whether released from a plane or thrown by a gorilla. The answer is D.

(C) A 0.2-kg banana is released from rest at the same time as a 20-kg box of bananas from the same height above horizontal ground. Which object strikes the ground first?

Solution. Look at the equations of free fall (Chapter 3). Observe that **mass** is not a factor in those equations. Neglecting air resistance, all objects fall with the same acceleration, **regardless of mass**. (Even allowing for air resistance, a banana and box of bananas released from a couple of meters above the ground will strike the ground at nearly the same time.) The banana and box of bananas both have an acceleration of $a_y = -9.81$ m/s^2. Since they have the same acceleration and descend the same height, and since they are released simultaneously, they reach the ground at the same time. The answer is: "They strike the ground at the same time."

(D) One bullet is released from rest and falls straight down at the same time as a second bullet with identical mass is shot horizontally to the right from the same height above horizontal ground. Which bullet strikes the ground first?

Solution. Note that gravity affects both bullets the same way: Both bullets have the **same acceleration**, $a_y = -9.81$ m/s^2, since both bullets are in **free fall**. Both bullets have $v_{y0} = 0$ (the shot bullet has a nonzero v_{x0}, but its v_{y0} is zero). The y-equations of projectile motion (Chapter 11) are identical to the equations of one-dimensional uniform acceleration (Chapter 3). So t will be the same for each. The answer is: "They strike the ground at the same time."

(E) A necklace dangles from the rearview mirror of a car. The car drives on a level road. Which way does the necklace lean when the car speeds up? slows down? travels with constant velocity? rounds a turn to the left with constant speed?

Solution. Apply Newton's first law, which involves the concept of **inertia**. Objects have **inertia**, which is a natural tendency to travel with constant velocity. The necklace resists changes to its velocity. The necklace resists acceleration.

- When the car speeds up, the necklace resists this increase in velocity. The necklace leans **backward** when the car speeds up (assuming the car isn't in reverse).
- When the car slows down, the necklace resists this decrease in velocity. The necklace leans **forward** when the car slows down (assuming the car isn't in reverse).
- When the car travels with constant velocity along a level road, the necklace has nothing to resist, so it leans **straight down**.
- When the car rounds a turn to the left with constant speed, the necklace wants to keep going in a straight line due to its inertia, so it leans to the driver's **right**. The necklace leans outward.

14 APPLICATIONS OF NEWTON'S SECOND LAW

Equations			
$\sum F_x = ma_x$, $\sum F_y = ma_y$	$W = mg$	$f_k = \mu_k N$, $f_s \le \mu_s N$

Symbol	Name	SI Units
m	mass	kg
mg	weight	N
F	force	N
N	normal force	N
f	friction force	N
f_s	force of static friction	N
f_k	force of kinetic friction	N
μ	coefficient of friction	unitless
μ_s	coefficient of static friction	unitless
μ_k	coefficient of kinetic friction	unitless
T	tension	N
D	drive force	N
L	lift force	N
P	another kind of push or a pull	N
a_x	x-component of acceleration	m/s^2
a_y	y-component of acceleration	m/s^2

Example 91. As shown below, a 600-kg helicopter flies straight upward with a lift force of 12,000-N. A 150-kg spy (00π) hangs onto a rope that is connected to the helicopter. A 50-kg physics student hangs onto a second rope, which is held by the spy. Determine the acceleration of the system and the tension in each cord.

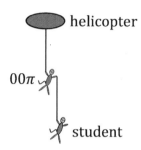

Solution. Draw and label a FBD (free-body diagram) for each object.
- Each object has **weight** ($m_1\vec{g}$, $m_2\vec{g}$, and $m_3\vec{g}$) pulling straight down.
- There are two pairs of **tension** ($\vec{T}_1$ and $\vec{T}_2$) forces: one pair in each rope. According to Newton's third law (Chapter 13), tension comes in pairs.
- A **lift force** ($\vec{L}$) pulls upward on the helicopter.
- Label $+x$ in the direction of the acceleration: straight up.

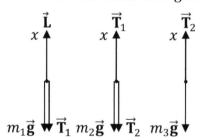

Apply Newton's second law to each object.
$$\sum F_{1x} = m_1 a_x \quad , \quad \sum F_{2x} = m_2 a_x \quad , \quad \sum F_{3x} = m_3 a_x$$
Rewrite each sum using the forces from the FBD's. Forces pulling upward are positive, while forces pulling downward are negative (since $+x$ is up).
$$L - m_1 g - T_1 = m_1 a_x \quad , \quad T_1 - m_2 g - T_2 = m_2 a_x \quad , \quad T_2 - m_3 g = m_3 a_x$$
Add all three equations together in order to cancel the tension forces. The sum of the left-hand sides equals the sum of the right-hand sides.
$$L - m_1 g - T_1 + T_1 - m_2 g - T_2 + T_2 - m_3 g = m_1 a_x + m_2 a_x + m_3 a_x$$
$$L - m_1 g - m_2 g - m_3 g = m_1 a_x + m_2 a_x + m_3 a_x$$
$$12,000 - 600(9.81) - 150(9.81) - 50(9.81) = 600 a_x + 150 a_x + 50 a_x$$
$$12,000 - 5886 - 1471.5 - 490.5 = 800 a_x$$
$$4152 = 800 a_x$$

Divide both sides by 800.
$$a_x = \frac{4152}{800} = 5.19 \text{ m/s}^2$$

Solve for tension in the equations from Newton's second law (the sums). Begin with the equation from the first sum (for the helicopter).

$$L - m_1 g - T_1 = m_1 a_x$$

Add T_1 to both sides.

$$L - m_1 g = T_1 + m_1 a_x$$

Subtract $m_1 a_x$ from both sides. Since we're solving for T_1, let's swap the left and right sides. Since $L - m_1 g - m_1 a_x = T_1$, it's also true that $T_1 = L - m_1 g - m_1 a_x$.

$$T_1 = L - m_1 g - m_1 a_x$$
$$T_1 = 12{,}000 - 600(9.81) - 600(5.19)$$
$$T_1 = 12{,}000 - 5886 - 3114$$
$$T_1 = 3000 \text{ N}$$

Now use the equation from the right sum (for the student).

$$T_2 - m_3 g = m_3 a_x$$

Add $m_3 g$ to both sides.

$$T_2 = m_3 a_x + m_3 g$$
$$T_2 = 50(5.19) + 50(9.81)$$
$$T_2 = 259.5 + 490.5$$
$$T_2 = 750 \text{ N}$$

The acceleration is $a_x = 5.19 \text{ m/s}^2$ and the tensions are $T_1 = 3000$ N and $T_2 = 750$ N.

Example 92. As illustrated below, a 30-kg monkey is connected to a 20-kg box of bananas by a cord that passes over a pulley. The coefficient of friction between the box and ground is 0.50. The system is initially at rest. Determine the acceleration of the system and the tension in the cord.

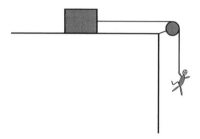

Solution. Draw and label a FBD for each object.

- Each object has **weight** ($m_1\vec{g}$ and $m_2\vec{g}$) pulling straight down.
- There is a pair of **tension** ($\vec{T}$) forces in the connecting cord. According to Newton's third law (Chapter 13), tension comes in pairs. Tension pulls the box to the right and pulls the monkey up (the monkey falls downward, yet tension pulls up: the tension reduces the acceleration with which the monkey falls).
- A **normal force** ($\vec{N}$) supports the box (perpendicular to the surface). Since the surface is horizontal, normal force points up.
- **Friction** ($\vec{f}$) acts opposite to the velocity of the box of bananas: to the left. (Although the box begins from rest, it will travel to the right once the motion starts.)
- Label $+x$ in the direction that each object accelerates: to the right for the box and down for the monkey. Think of the pulley as bending the x-axis.
- Label $+y$ perpendicular to x. For the box of bananas, $+y$ is up.

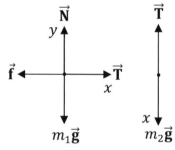

The monkey and box have the same acceleration because they are connected by a cord (which we will assume doesn't stretch). We choose x to bend around the pulley so that we can use the same symbol (a_x) for the acceleration of both objects. Apply Newton's second law to each object. We don't need $\sum F_{2y}$ because there are no forces pulling on the monkey with a horizontal component.

$$\sum F_{1x} = m_1 a_x \quad , \quad \sum F_{1y} = m_1 a_y \quad , \quad \sum F_{2x} = m_2 a_x$$

Rewrite each sum using the forces from the FBD's. Note that $a_y = 0$ because the box doesn't accelerate vertically: The box has a_x, but not a_y.

$$T - f = m_1 a_x \quad , \quad N - m_1 g = 0 \quad , \quad m_2 g - T = m_2 a_x$$

Compare the sums to the FBD's. Tension and friction pull horizontally on the box, with tension positive (right) and friction negative (left): $T - f = m_1 a_x$. Normal force and weight pull vertically on the box, with normal force positive (up) and weight negative (down): $N - m_1 g = m_1 a_y$. Weight and tension pull vertically on the monkey, with weight positive (down) and tension negative (up): $m_2 g - T = m_2 a_x$. For the monkey, down is positive while up is negative because $+x$ is down.

When there is friction in a problem, first solve for normal force in the y-sum.

$$N - m_1 g = 0$$
$$N = m_1 g = 20(9.81) = 196 \text{ N}$$

Then use the equation for friction.

$$f = \mu N = 0.5(196) = 98 \text{ N}$$

When there is a connecting cord in a problem, add the x-equations together in order to cancel tension. The sum of the left-hand sides equals the sum of the right-hand sides.

$$T - f + m_2 g - T = m_1 a_x + m_2 a_x$$
$$-f + m_2 g = m_1 a_x + m_2 a_x$$
$$-98 + 30(9.81) = 20 a_x + 30 a_x$$
$$-98 + 294.3 = 50 a_x$$
$$196.3 = 50 a_x$$
$$a_x = \frac{196.3}{50} = 3.93 \text{ m/s}^2$$

Plug acceleration into one of the original equations from the x-sums in order to solve for the tension.

$$T - f = m_1 a_x$$
$$T = f + m_1 a_x$$
$$T = 98 + 20(3.93)$$
$$T = 98 + 78.6$$
$$T = 177 \text{ N}$$

The acceleration is $a_x = 3.93 \text{ m/s}^2$ and the tension is $T = 177 \text{ N}$.

Example 93. Two boxes of bananas connected by a cord are pulled by a monkey with a force of $300\sqrt{3}$ N as illustrated below. The left box of bananas has a mass of $10\sqrt{3}$ kg and the right box of bananas has a mass of $30\sqrt{3}$ kg. The coefficient of friction between the boxes of bananas and the horizontal is $\frac{\sqrt{3}}{5}$. The system is initially at rest. Determine the acceleration of the system and the tension in the cord.

Solution. Draw and label a FBD for each box.

- Each object has **weight** ($m_1\vec{g}$ and $m_2\vec{g}$) pulling straight down.
- There is a pair of **tension** ($\vec{T}$) forces in the connecting cord. According to Newton's third law (Chapter 13), tension comes in equal-and-opposite pairs.
- **Normal forces** ($\vec{N}_1$ and $\vec{N}_2$) support the boxes (perpendicular to the surface). Since the surface is horizontal, normal force points up.
- Two different friction forces ($\vec{f}_1$ and $\vec{f}_2$) pull backwards on the two boxes.
- The monkey's pull ($\vec{P}$) acts on the right box.
- Label $+x$ in the direction that each object accelerates: to the right.
- Label $+y$ perpendicular to x. Since x is horizontal, $+y$ is up.

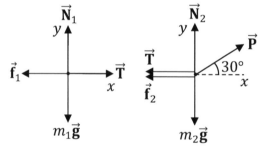

Apply Newton's second law to each box.
$$\sum F_{1x} = m_1 a_x \quad , \quad \sum F_{1y} = m_1 a_y \quad , \quad \sum F_{2x} = m_2 a_x \quad , \quad \sum F_{2y} = m_2 a_y$$
Rewrite each sum using the forces from the FBD's. Note that $a_y = 0$ because the boxes don't accelerate vertically: The boxes have a_x, but not a_y.
$$T - f_1 = m_1 a_x \ , \ N_1 - m_1 g = 0 \ , \ P \cos 30° - T - f_2 = m_2 a_x \ , \ N_2 + P \sin 30° - m_2 g = 0$$
Since $\vec{P}$ doesn't lie on an axis, it goes in both the $\sum F_{2x}$ and $\sum F_{2y}$ sums with trig. The **components** of $\vec{P}$ are $P_x = P \cos 30°$ and $P_y = P \sin 30°$. The horizontal forces go in the x-sums (right is positive and left is negative) and the vertical forces go in the y-sums (up is positive and down is negative).

When there is friction in a problem, first solve for normal force in the y-sum. In this problem, there are two different normal forces to solve for.

$$N_1 - m_1 g = 0 \quad , \quad N_2 + P \sin 30° - m_2 g = 0$$
$$N_1 = m_1 g \quad , \quad N_2 = m_2 g - P \sin 30°$$
$$N_1 = 10\sqrt{3}(9.81) \quad , \quad N_2 = 30\sqrt{3}(9.81) - 300\sqrt{3}\left(\frac{1}{2}\right)$$

We will round 9.81 to ≈ 10 because the arithmetic is instructive. (If you prefer, you can use 9.81 and follow the math with a calculator. This will let you work with decimals instead of fractions and squareroots. You will get approximately the same answer, provided that you do the computations correctly.)

$$N_1 \approx 10\sqrt{3}(10) \quad , \quad N_2 \approx 30\sqrt{3}(10) - 300\sqrt{3}\left(\frac{1}{2}\right)$$
$$N_1 \approx 100\sqrt{3} \text{ N} \quad , \quad N_2 \approx 300\sqrt{3} - 150\sqrt{3} = 150\sqrt{3} \text{ N}$$

Now use the equation for friction.

$$f_1 = \mu N_1 \approx \frac{\sqrt{3}}{5}(100\sqrt{3}) = \frac{100}{5}\sqrt{3}\sqrt{3} = 20(3) = 60 \text{ N}$$
$$f_2 = \mu N_2 \approx \frac{\sqrt{3}}{5}(50\sqrt{3}) = \frac{150}{5}\sqrt{3}\sqrt{3} = 30(3) = 90 \text{ N}$$

When there is a connecting cord in a problem, add the x-equations together in order to cancel tension. The sum of the left-hand sides equals the sum of the right-hand sides.

$$T - f_1 + P \cos 30° - T - f_2 = m_1 a_x + m_2 a_x$$
$$-f_1 + P \cos 30° - f_2 = m_1 a_x + m_2 a_x$$
$$-60 + 300\sqrt{3}\left(\frac{\sqrt{3}}{2}\right) - 90 \approx 10\sqrt{3}a_x + 30\sqrt{3}a_x$$
$$-60 + \frac{300}{2}\sqrt{3}\sqrt{3} - 90 \approx 40\sqrt{3}a_x$$
$$-60 + 150(3) - 90 \approx 40\sqrt{3}a_x$$
$$-60 + 450 - 90 \approx 40\sqrt{3}a_x$$
$$300 \approx 40\sqrt{3}a_x$$
$$a_x \approx \frac{300}{40\sqrt{3}} = \frac{300}{40\sqrt{3}}\frac{\sqrt{3}}{\sqrt{3}} = \frac{300\sqrt{3}}{40(3)} = \frac{100\sqrt{3}}{40} = \frac{5\sqrt{3}}{2} \text{ m/s}^2$$

We multiplied by $\frac{\sqrt{3}}{\sqrt{3}}$ to **rationalize the denominator** in order to express our answer in **standard form**. Note that $\sqrt{3}\sqrt{3} = 3$ and $\frac{100}{40} = \frac{100 \div 20}{40 \div 20} = \frac{5}{2}$. Plug acceleration into one of the original equations from the x-sums in order to solve for the tension.

$$T - f_1 = m_1 a_x$$
$$T = f_1 + m_1 a_x$$
$$T \approx 60 + 10\sqrt{3}\left(\frac{5\sqrt{3}}{2}\right) = 60 + \frac{10(5)}{2}\sqrt{3}\sqrt{3} = 60 + 25(3) = 60 + 75 = 135 \text{ N}$$

The acceleration is $a_x = \frac{5\sqrt{3}}{2}$ m/s^2 and the tension is $T = 135$ N. (If you don't round gravity, the acceleration is $a_x = 4.4$ m/s^2.)

Example 94. A box of bananas slides down an inclined plane that makes an angle of 30° with the horizontal. The coefficient of friction between the box and the incline is $\frac{\sqrt{3}}{5}$. Determine the acceleration of the box of bananas.

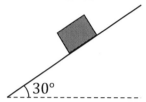

Solution. Draw and label a FBD for the box of bananas.

- **Weight** ($m\vec{g}$) pulls straight down.
- **Normal force** ($\vec{N}$) pushes perpendicular to the surface. Since the surface is inclined, normal force is at an angle (see below) in order to be perpendicular to the incline.
- Friction ($\vec{f}$) acts opposite to the velocity of the box of bananas: up the incline.
- Label $+x$ in the direction that the box accelerates: down the incline. (Note that x is **not** horizontal in this problem.)
- Since y must be perpendicular to x, we choose $+y$ to be along the normal force. (Note that y is **not** vertical in this problem). The reason for this is that this choice of coordinates makes $a_y = 0$, since the object won't accelerate perpendicular to the incline. The benefit is that it makes the math much simpler.

Apply Newton's second law to the box of bananas. Since weight doesn't lie on an axis, it appears in both the x- and y-sums with trig. Study the diagram below.

The x-component of weight has a sine since x is opposite to 30°, while the y-component of weight has a cosine since y is adjacent to 30° (see the figure above). Normal force only appears in the y-sum because it lies on the y-axis. Friction only appears in the x-sum because it lies on the negative x-axis. Friction is negative because it is opposite to $+x$.

$$\sum F_x = ma_x \quad , \quad \sum F_y = ma_y$$
$$mg\sin 30° - f = ma_x \quad , \quad N - mg\cos 30° = 0$$

Note that $a_y = 0$ since the box won't accelerate perpendicular to the incline (it accelerates down the incline, along x, and does not accelerate along y, which is perpendicular to the incline). When there is friction in a problem, first solve for normal force in the y-sum.

$$N - mg \cos 30° = 0$$

$$N = mg \cos 30° = \frac{mg\sqrt{3}}{2}$$

Since we don't know the mass of the box of bananas, we won't get a numerical value for normal force, but that doesn't matter. We will simply work with the algebraic expression $N = \frac{mg\sqrt{3}}{2}$. Substitute this expression for normal force into the equation for friction.

$$f = \mu N = \frac{\sqrt{3}}{5}\left(\frac{mg\sqrt{3}}{2}\right) = \frac{\sqrt{3}\sqrt{3}}{5(2)}mg = \frac{3mg}{10}$$

Again, we can't obtain a number for friction force because we don't know the mass of the box, so we will just work with the expression $f = \frac{3mg}{10}$. Plug this expression for friction force into the equation from the x-sum.

$$mg \sin 30° - f = ma_x$$

$$mg \sin 30° - \frac{3mg}{10} = ma_x$$

$$\frac{mg}{2} - \frac{3mg}{10} = ma_x$$

Now the mass cancels out. Since every term has mass, divide both sides by m.

$$\frac{g}{2} - \frac{3g}{10} = a_x$$

$$a_x = \frac{9.81}{2} - \frac{3(9.81)}{10}$$

$$a_x = 4.905 - 2.943$$

$$a_x = 1.96 \text{ m/s}^2$$

Example 95. As illustrated below, a 40-kg box of bananas on a frictionless incline is connected to a 60-kg monkey by a cord that passes over a pulley. The system is initially at rest. Determine the acceleration of each object and the tension in the connecting cord.

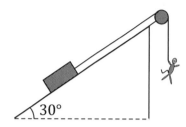

Solution. Draw and label a FBD for each object.

- Each object has **weight** ($m_1\vec{\mathbf{g}}$ and $m_2\vec{\mathbf{g}}$) pulling straight down.
- There is a pair of **tension** ($\vec{\mathbf{T}}$) forces in the connecting cord. According to Newton's third law (Chapter 13), tension comes in pairs. Tension pulls the box up the incline and pulls the monkey up (the monkey falls downward, yet tension pulls up: the tension reduces the acceleration with which the monkey falls).
- **Normal force** ($\vec{\mathbf{N}}$) pushes perpendicular to the surface. Since the surface is inclined, normal force is at an angle (see below) in order to be perpendicular to the incline.
- Label $+x$ in the direction that each object accelerates: up the incline for the box and down for the monkey. Think of the pulley as bending the x-axis.
- Label $+y$ perpendicular to x. For the box of bananas, $+y$ is along the normal force.

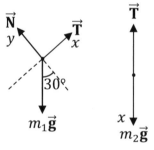

The monkey and box have the same acceleration because they are connected by a cord (which we will assume doesn't stretch). We choose x to bend around the pulley so that we can use the same symbol (a_x) for the acceleration of both objects. Apply Newton's second law to each object. We don't need $\sum F_{2y}$ because there are no forces pulling on the monkey with a horizontal component.

$$\sum F_{1x} = m_1 a_x \quad , \quad \sum F_{1y} = m_1 a_y \quad , \quad \sum F_{2x} = m_2 a_x$$

Rewrite each sum using the forces from the FBD's. Note that $a_y = 0$ because the box doesn't accelerate vertically: The box has a_x, but not a_y.

$$T - m_1 g \sin 30° = m_1 a_x \quad , \quad N - m_1 g \cos 30° = 0 \quad , \quad m_2 g - T = m_2 a_x$$

The previous example explains why there is $m_1 g \sin 30°$ in the x-sum and $m_1 g \cos 30°$ in the y-sum. Example 92 explains how the other forces fit into these sums.

When there is a connecting cord in a problem, add the x-equations together in order to cancel tension. The sum of the left-hand sides equals the sum of the right-hand sides.

$$T - m_1 g \sin 30° + m_2 g - T = m_1 a_x + m_2 a_x$$
$$-(40)(9.81)\frac{1}{2} + (60)(9.81) = 40a_x + 60a_x$$
$$-196.2 + 588.6 = 100a_x$$
$$392.4 = 100a_x$$
$$a_x = \frac{392.4}{100} = 3.93 \text{ m/s}^2$$

Plug acceleration into one of the original equations from the x-sums in order to solve for the tension.

$$T - m_1 g \sin 30° = m_1 a_x$$
$$T = m_1 g \sin 30° + m_1 a_x$$
$$T = (40)(9.81)\frac{1}{2} + (40)(3.93)$$
$$T = 196.2 + 157.2$$
$$T = 353 \text{ N}$$

The acceleration is $a_x = 3.93$ m/s^2 and the tension is $T = 353$ N.

Example 96. As illustrated below, a 40-kg chest filled with bananas is on top of an 80-kg chest filled with coconuts. The coefficient of static friction between the chests is $\frac{1}{4}$, but the ground is frictionless. A monkey exerts a 200-N horizontal force on the top chest. Determine the acceleration of the top chest and the tension in the cord.

Solution. Draw and label a FBD for each box. We call the top box object 1 (right figure) and the bottom box object 2 (left figure).

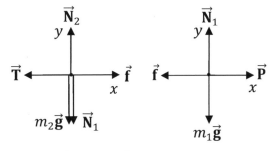

- Each object has **weight** ($m_1\vec{g}$ and $m_2\vec{g}$) pulling straight down.
- The monkey's pull ($\vec{P}$) acts on the top box.
- Tension ($\vec{T}$) in the cord pulls the bottom box to the left.

- **Normal forces** ($\vec{N}_1$ and $\vec{N}_2$) support the boxes (perpendicular to the surface). There is also an equal and opposite normal force $\vec{N}_1$ pushing downward on the bottom box. This follows from **Newton's third law**. Since the bottom box pushes upward on the top box with a force $\vec{N}_1$, the top box must push downward on the bottom box with an equal and opposite force.

- Friction force ($\vec{f}$) pulls to the left on the top box (resisting the 200-N pull). There is an equal and opposite friction force ($\vec{f}$) pulling the bottom box to the right due to **Newton's third law**. This friction force creates tension in the cord.

- Label $+x$ in the direction that the top box accelerates: to the right.

- Label $+y$ perpendicular to x. Since x is horizontal, $+y$ is up.

Apply Newton's second law to each box.

$$\sum F_{1x} = m_1 a_{1x} \quad , \quad \sum F_{1y} = m_1 a_y \quad , \quad \sum F_{2x} = m_2 a_{2x} \quad , \quad \sum F_{2y} = m_2 a_y$$

Rewrite each sum using the forces from the FBD's. Note that $a_y = 0$ because the boxes don't accelerate vertically. Also note that $a_{2x} = 0$, but a_{1x} does not.

$$P - f = m_1 a_{1x} \quad , \quad N_1 - m_1 g = 0 \quad , \quad f - T = 0 \quad , \quad N_2 - N_1 - m_2 g = 0$$

When there is friction in a problem, first solve for normal force in the y-sum. In this problem, there are two different normal forces to solve for.

$$N_1 - m_1 g = 0 \quad , \quad N_2 - N_1 - m_2 g = 0$$
$$N_1 = m_1 g \quad , \quad N_2 = N_1 + m_2 g$$
$$N_1 = 40(9.81) \quad , \quad N_2 = 40(9.81) + 80(9.81)$$
$$N_1 = 392.4 \text{ N} \quad , \quad N_2 = 392.4 + 784.8$$
$$N_1 = 392.4 \text{ N} \quad , \quad N_2 = 1177.2 \text{ N}$$

Now use the equation for friction. If you look at the FBD for the top box (which we called box 1), you will see that there is no ambiguity: The correct normal force to use is N_1.

$$f = \mu N_1 = \frac{1}{4}(392.4) = 98.1 \text{ N}$$

Now we can solve for acceleration using an equation from the x-sums.

$$P - f = m_1 a_{1x}$$
$$200 - 98.1 = 40 a_{1x}$$
$$101.9 = 40 a_{1x}$$
$$a_{1x} = \frac{101.9}{40} = 2.55 \text{ m/s}^2$$

Solve for tension using the equation from the other x-sum.

$$f - T = 0$$
$$T = f = 98.1 \text{ N}$$

The acceleration is $a_{1x} = 2.55 \text{ m/s}^2$ and the tension is $T = 98.1 \text{ N}$. (If you don't round gravity, $a_{1x} = \frac{5}{2}$ m/s^2 and $T = 100$ N.)

15 HOOKE'S LAW

Equation	Symbol	Name	SI Units
$F_r = -k\Delta x$	k	spring constant	N/m (or kg/s^2)
	Δx	displacement from equilibrium	m
	F_r	restoring force	N

Example 97. As illustrated below, a spring with a spring constant of 45 N/m on a horizontal surface has its left end attached to a wall while its right end is attached to a 3.0-kg box of bananas. The coefficient of friction between the box and the horizontal surface is $\frac{1}{4}$. What is the acceleration of the spring when it is stretched 50 cm to the right of its equilibrium position while moving to the left (heading back to equilibrium)?

Solution. Draw and label a FBD for the box of bananas.

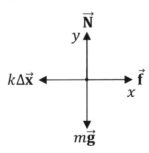

- **Weight** ($m\vec{\mathbf{g}}$) pulls straight down.
- A **normal force** ($\vec{\mathbf{N}}$) supports the box (perpendicular to the surface). Since the surface is horizontal, normal force points up.
- **Friction** ($\vec{\mathbf{f}}$) acts opposite to the velocity of the box of bananas: to the right (for the situation described in the problem, where the box is presently moving to the left).

107

- The spring exerts a **restoring force** ($k\Delta\vec{x}$) on the box, where k is the spring constant and Δx is the displacement from equilibrium. The restoring force is directed to the left (since the spring is stretched from equilibrium in the situation described).
- We choose $+x$ to point to the right and $+y$ to point up.

Apply Newton's second law to the box. The restoring force and friction pull horizontally on the box, with friction positive (right) and restoring force negative (left) since we setup our coordinate system with $+x$ pointing to the right. Normal force and weight pull vertically on the box, with normal force positive (up) and weight negative (down).

$$\sum F_x = ma_x \quad , \quad \sum F_y = ma_y$$

Rewrite each sum using the forces from the FBD's. Note that $a_y = 0$ because the box doesn't accelerate vertically: The box has a_x, but not a_y.

$$-k|\Delta x| + f = ma_x \quad , \quad N - mg = 0$$

We put absolute values around the displacement from equilibrium, $|\Delta x|$, because we already used the minus sign from Hooke's law ($F_r = -k\Delta x$) when we drew the restoring force toward equilibrium in our FBD. That's why we wrote $-k|\Delta x|$ in the above equation. When there is friction in a problem, first solve for normal force in the y-sum.

$$N - mg = 0$$
$$N = mg = 3(9.81) = 29.4 \text{ N}$$

Then use the equation for friction.

$$f = \mu N = \frac{1}{4}(29.4) = 7.35 \text{ N}$$

Note that the spring constant is $k = 45$ N/m and the displacement from equilibrium is $|\Delta x| = 50$ cm. Since the spring constant is in N/m, we must convert $|\Delta x|$ from centimeters (cm) to meters (m):

$$|\Delta x| = 50 \text{ cm} = 50 \text{ cm} \frac{1 \text{ m}}{100 \text{ cm}} = \frac{1}{2} \text{ m}$$

Now use the equation from the x-sum to solve for acceleration.

$$-k|\Delta x| + f = ma_x$$
$$-45\left(\frac{1}{2}\right) + 7.35 = 3a_x$$
$$-22.5 + 7.35 = 3a_x$$
$$-15.15 = 3a_x$$
$$a_x = -\frac{15.15}{3} = -5.1 \text{ m/s}^2$$

The acceleration is $a_x = -5.1$ m/s^2. The minus sign means that the box is presently accelerating to the **left** (since we chose $+x$ to be right).

Example 98. As illustrated below, a spring is stretched on a frictionless incline by fixing its upper end and connecting a 36-kg box of bananas to its lower end. When the spring is in its new inclined equilibrium position, it is 4.0 m longer than when lying in its natural horizontal equilibrium position. Determine the spring constant.

Solution. Draw and label a FBD for the box of bananas.

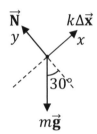

- **Weight** ($m\vec{g}$) pulls straight down.
- **Normal force** ($\vec{N}$) pushes perpendicular to the surface. Since the surface is inclined, normal force is at an angle (see above) in order to be perpendicular to the incline.
- The spring exerts a **restoring force** ($k\Delta\vec{x}$) on the box, where k is the spring constant and Δx is the displacement from equilibrium. The restoring force is directed up the incline (toward its original equilibrium position).
- Like we did in Example 95, label $+x$ up the incline. (Note that x is **not** horizontal in this problem.)
- Since y must be perpendicular to x, we choose $+y$ to be along the normal force. (Note that y is **not** vertical in this problem).

Apply Newton's second law to the box of bananas. Example 94 explains why there is $mg \sin 30°$ in the x-sum and $mg \cos 30°$ in the y-sum.

$$\sum F_x = ma_x \quad , \quad \sum F_y = ma_y$$

$$k|\Delta x| - mg \sin 30° = 0 \quad , \quad N - mg \cos 30° = 0$$

As in the previous example, we placed absolute values around the displacement from equilibrium, $|\Delta x|$, because we already applied the sign in Hooke's law ($F_r = -k\Delta x$) to draw the restoring force toward the box's original equilibrium position. Note that both $a_x = 0$ and $a_y = 0$ because the box of bananas is in equilibrium (it isn't accelerating at all).

We can solve for the spring constant in the equation from the x-sum by setting $a_x = 0$ (since the box isn't in motion, but is resting in its new equilibrium position).

$$k|\Delta x| - mg \sin 30° = 0$$
$$k|\Delta x| = mg \sin 30°$$

Plug in numbers. Note that the displacement from equilibrium is $|\Delta x| = 4.0$ m.

$$k(4) = 36(9.81)\frac{1}{2}$$
$$4k = 176.6$$
$$k = \frac{176.6}{4}$$
$$k = 44 \text{ N/m}$$

The spring constant is $k = 44$ N/m.

16 UNIFORM CIRCULAR MOTION

Equations			
$a_c = \dfrac{v^2}{R}$	$v = R\omega$	$v = \dfrac{s}{t}$	$v = \dfrac{2\pi R}{T}$
$\omega = \dfrac{\theta}{t}$	$\omega = \dfrac{2\pi}{T}$	$\omega = 2\pi f$	$f = \dfrac{1}{T}$
$s = R\theta$	$C = 2\pi R$	$D = 2R$	$\sum F_{in} = ma_c$

Symbol	Name	SI Units
a_c	centripetal acceleration	m/s^2
v	speed	m/s
ω	angular speed	rad/s
s	total distance traveled (arc length)	m
C	circumference	m
R	radius	m
D	diameter	m
θ	angle	rad
t	time	s
T	period	s
f	frequency	Hz
m	mass	kg
F	force	N

Note: The symbol for angular speed (ω) is the lowercase Greek letter omega.

Example 99. A monkey skates in a 12-m diameter circle with a constant angular speed of $\frac{1}{2}$ rad/s for a total time of 4.0 minutes.

(A) What is the monkey's speed?

Solution. Begin by making a list of the known quantities.

- The diameter of the circle is $D = 12$ m. Therefore the radius is $R = 6$ m.
- The angular speed of the monkey is $\omega = \frac{1}{2}$ rad/s.
- The total time spent traveling is $t = 4.0$ min. $= 240$ s.

Note that we converted the time to seconds (s) and found the radius as follows.

$$t = 4.0 \text{ min.} = 4.0 \text{ min.}\frac{60 \text{ s}}{1 \text{ min.}} = 240 \text{ s}$$

$$R = \frac{D}{2} = \frac{12}{2} = 6 \text{ m}$$

Since know the radius ($R = 6$ m) and angular speed ($\omega = \frac{1}{2}$ rad/s), we can use the equation $v = R\omega$ to find the speed.

$$v = R\omega = (6)\left(\frac{1}{2}\right) = 3.0 \text{ m/s}$$

The monkey's speed is $v = 3.0$ m/s. Note that speed (v) and angular speed (ω) are two different quantities.

(B) What is the monkey's acceleration?

Solution. Use the speed ($v = 3.0$ m/s) from part (A) along with the radius ($R = 6$ m).

$$a_c = \frac{v^2}{R} = \frac{(3)^2}{6} = \frac{9}{6} = \frac{3}{2} \text{ m/s}^2$$

The monkey's acceleration is $a_c = \frac{3}{2}$ m/s^2, which is the same as 1.5 m/s^2. Note that the acceleration is **not** zero because the **direction** of the monkey's velocity is changing. This type of acceleration is called **centripetal acceleration**. (Recall that acceleration, in general, is the instantaneous rate at which velocity changes.)

(C) How far does the monkey travel?

Solution. Since we found the speed ($v = 3.0$ m/s) in part (A) and since we were given the total time ($t = 240$ s), we can use the equation $s = vt$.

$$s = vt = (3)(240) = 720 \text{ m}$$

The total distance traveled is $s = 720$ m. (Since the speed is in **meters per second**, it would be a mistake to leave the time in **minutes**.)

(D) What is the period of revolution?

Solution. Use the angular speed ($\omega = \frac{1}{2}$ rad/s) and the equation $\omega = \frac{2\pi}{T}$.

$$\omega = \frac{2\pi}{T}$$

Multiply both sides of the equation by the period (T).

$$\omega T = 2\pi$$

Divide both sides of the equation by angular speed (ω).

$$T = \frac{2\pi}{\omega} = \frac{2\pi}{\frac{1}{2}}$$

To divide by a fraction, multiply by its **reciprocal**. Note that the reciprocal of $\frac{1}{2}$ is 2.

$$T = \frac{2\pi}{\frac{1}{2}} = 2\pi(2) = 4\pi \text{ s}$$

The period of revolution is $T = 4\pi$ s, which works out to approximately $T \approx 13$ s if you use a calculator. (Note that lowercase t represents the total time, whereas uppercase T, called the period, is the time it takes to go around the circle exactly once.)

(E) How many revolutions does the monkey complete?

Solution. One way to answer this question is to solve for the angle θ and then convert the answer from radians to revolutions. We can use the total distance traveled ($s = 720$ m) from part (C) and the radius ($R = 6$ m) to determine the angle θ in radians.

$$s = R\theta$$

To solve for θ, divide both sides by R.

$$\theta = \frac{s}{R} = \frac{720}{6} = 120 \text{ rad}$$

In order to answer the question, we must convert θ from radians to revolutions. The conversion factor is 1 rev $= 2\pi$ rad.

$$\theta = 120 \text{ rad} = 120 \text{ rad}\frac{1 \text{ rev}}{2\pi \text{ rad}} = \frac{60}{\pi} \text{ rev}$$

The monkey completes $\theta = \frac{60}{\pi}$ revolutions, which equates to 19 revolutions.

Example 100. A monkey runs in circle with a constant speed of 4.0 m/s and a period of 8π s, completing $\frac{20}{\pi}$ revolutions.

(A) What is the monkey's angular speed?

Solution. Begin by making a list of the known quantities.

- The speed of the monkey is $v = 4.0$ m/s.
- The period of revolution is $T = 8\pi$ s.
- The number of revolutions can be converted into radians. $\theta = \frac{20}{\pi}$ rev $= 40$ rad.

We converted the angle from revolutions to radians as follows.

$$\theta = \frac{20}{\pi} \text{ rev} = \frac{20}{\pi} \text{ rev}\frac{2\pi \text{ rad}}{1 \text{ rev.}} = \frac{20(2\pi)}{\pi} \text{ rad} = 40 \text{ rad}$$

To find the angular speed (ω), use the period ($T = 8\pi$ s) and the equation $\omega = \frac{2\pi}{T}$.

$$\omega = \frac{2\pi}{T} = \frac{2\pi}{8\pi} = \frac{1}{4} \text{ rad/s}$$

The angular speed of the monkey is $\omega = \frac{1}{4}$ rad/s, which is the same as 0.25 rad/s.

(B) What is the radius of the circle?

Solution. We can use the angular speed ($\omega = \frac{1}{4}$ rad/s) from part (A) along with the speed ($v = 4.0$ m/s) to find the radius.

$$v = R\omega$$

Divide both sides of the equation by the angular speed (ω).

$$R = \frac{v}{\omega} = \frac{4}{1/4}$$

To divide by a fraction, multiply by its **reciprocal**. Note that the reciprocal of $\frac{1}{4}$ is 4.

$$R = \frac{4}{1/4} = 4(4) = 16 \text{ m}$$

The radius of the circle is $R = 16$ m.

(C) What is the monkey's acceleration?

Solution. Use the radius ($R = 16$ m) from part (B) along with the speed ($v = 4.0$ m/s).

$$a_c = \frac{v^2}{R} = \frac{(4)^2}{16} = \frac{16}{16} = 1.0 \text{ m/s}^2$$

The monkey's acceleration is $a_c = 1.0$ m/s².

(D) How far does the monkey travel?

Solution. Use the radius ($R = 16$ m) from part (B) along with the given angle ($\theta = 40$ rad).

$$s = R\theta = (16)(40) = 640 \text{ m}$$

The total distance traveled is $s = 640$ m. Note that θ must be in **radians** (not revolutions).

(E) What is the frequency?

Solution. Frequency (f) is the reciprocal of the period ($T = 8\pi$ s).

$$f = \frac{1}{T} = \frac{1}{8\pi} \text{ Hz}$$

The frequency is $f = \frac{1}{8\pi}$ Hz, which equates to 0.040 Hz.

17 UNIFORM CIRCULAR MOTION WITH NEWTON'S SECOND LAW

Equations			
$\sum F_{in} = ma_c$	$\sum F_{tan} = 0$	$\sum F_z = 0$	$a_c = \dfrac{v^2}{R}$

Symbol	Name	Units
a_c	centripetal acceleration	m/s^2
v	speed	m/s
R	radius	m
θ	angle	°
m	mass	kg
F	force	N

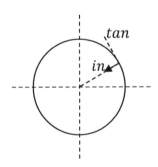

Coordinates		
in	inward	toward the center of the circle
tan	tangential	along a line that is tangent to the circle
z	axial	perpendicular to the plane of the circle

Example 101. An amusement park ride has a $48\sqrt{3}$-m diameter horizontal disc high above the ground. Several swings are suspended from the disc near its edge using $32\sqrt{3}$-m long chains. As the disc spins, the swings extend outward at an angle of $30°$ from the vertical, as illustrated below. A 25-kg monkey sitting in one of the swings travels in a horizontal circle.

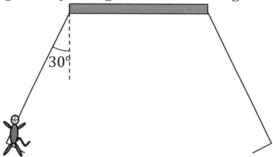

(A) What is the tension in the chain of the monkey's swing, which has a 5.0-kg seat?
Solution. Draw and label a FBD for the monkey.

- **Weight** ($m\vec{g}$) pulls straight down.
- **Tension** ($\vec{T}$) pulls along the chain.

In a circular motion problem, **don't** work with x- and y-coordinates. Instead, work with inward (in), tangential (tan), and the direction perpendicular to the plane of the circle (z).

- The inward (in) direction is toward the center of the circle: to the right for the position of the monkey shown above.
- The tangential (tan) direction is along a tangent: coming out of the page for the position shown. (Visualize the horizontal circle to help see this.)
- The z-direction is perpendicular to the plane of the circle: it points up.

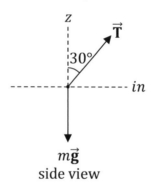

side view

Apply Newton's second law to the monkey. For tension, the inward component receives a sine because it is opposite to $30°$, while the z-component receives a cosine because z is adjacent to $30°$. Weight appears only in the z-sum since it lies on the negative z-axis. In circular motion, the acceleration is **centripetal** (a_c). The word "centripetal" means toward the center, so the sum of the inward components of the forces equals ma_c. The sum of the z-components of the forces is zero because the monkey doesn't accelerate up or down.

$$\sum F_{in} = ma_c \quad , \quad \sum F_z = 0$$
$$T \sin 30° = ma_c \quad , \quad T \cos 30° - mg = 0$$

We can solve for tension in the equation from the z-sum.
$$T \cos 30° - mg = 0$$
Add weight to both sides of the equation.
$$T \cos 30° = mg$$
Divide both sides of the equation by $\cos 30°$.
$$T = \frac{mg}{\cos 30°}$$
The mass equals the mass of the monkey **plus** the seat: $m = 25$ kg $+ 5$ kg $= 30$ kg.
$$T = \frac{(30)(9.81)}{\frac{\sqrt{3}}{2}}$$
To divide by a fraction, multiply by its **reciprocal**. Note that the reciprocal of $\frac{\sqrt{3}}{2}$ is $\frac{2}{\sqrt{3}}$.
$$T = \frac{(30)(9.81)}{\frac{\sqrt{3}}{2}} = (30)(9.81)\frac{2}{\sqrt{3}}$$
For this problem, we will round 9.81 to 10 because the arithmetic is instructive.
$$T \approx (30)(10)\frac{2}{\sqrt{3}} = \frac{600}{\sqrt{3}} \text{ N}$$
Note that this answer is not in **standard form**. In order to express the answer in standard form, we need to **rationalize the denominator**. Multiply the numerator and denominator by $\sqrt{3}$ in order to rationalize the denominator.
$$T \approx \frac{600}{\sqrt{3}} = \frac{600}{\sqrt{3}}\frac{\sqrt{3}}{\sqrt{3}} = \frac{600\sqrt{3}}{3} = 200\sqrt{3} \text{ N}$$
The tension is $T \approx 200\sqrt{3}$ N. If you don't round 9.81 to 10, the answer is $T = 340$ N.
(B) What is the monkey's speed?
Solution. Now that we know the tension, we can solve for acceleration in the inward sum.
$$T \sin 30° = ma_c$$
Divide both sides of the equation by mass.
$$a_c = \frac{T \sin 30°}{m} \approx \frac{200\sqrt{3}\left(\frac{1}{2}\right)}{30} = \frac{100\sqrt{3}}{30} = \frac{10\sqrt{3}}{3} \text{ m/s}^2$$
If you don't round gravity from 9.81 to 10, the tension is 340 N and the acceleration is 5.67 m/s^2. Use the formula for centripetal acceleration to solve for the speed.
$$a_c = \frac{v^2}{R}$$
Multiply both sides of the equation by the radius.
$$a_c R = v^2$$
Squareroot both sides of the equation.
$$v = \sqrt{a_c R}$$

Examine the figure below to see that the radius of the horizontal circle of the monkey's motion (labeled as R) equals the radius of the disc (R_{disc}) plus the bottom part of the right triangle formed by the chains (of length L). Note that the bottom part of that right triangle has a sine function because that side is opposite to the 30° angle.

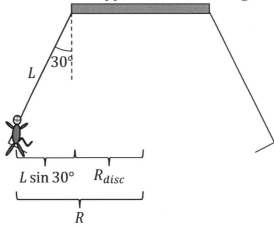

The radius of the circle of the monkey's motion is:

$$R = R_{disc} + L\sin 30° = \frac{D_{disc}}{2} + \frac{L}{2} = \frac{48\sqrt{3}}{2} + \frac{32\sqrt{3}}{2} = 24\sqrt{3} + 16\sqrt{3} = 40\sqrt{3} \text{ m}$$

Plug numbers into the previous equation for speed.

$$v = \sqrt{a_c R} \approx \sqrt{\left(\frac{10\sqrt{3}}{3}\right)(40\sqrt{3})} = \sqrt{\frac{(10)(40)}{3}\sqrt{3}\sqrt{3}} = \sqrt{\frac{(400)(3)}{3}} = \sqrt{400} = 20 \text{ m/s}$$

Note that $\sqrt{3}\sqrt{3} = 3$. The speed of the monkey is $v = 20$ m/s.

***

Example 102. A monkey grabs a 500-g mouse by the tail and whirls the mouse in a vertical circle with a constant speed of 4.0 m/s. The radius of the circle is 200 cm.

(A) What is the acceleration of the mouse?

Solution. Use the equation for centripetal acceleration. Since the speed is in meters per second, first convert the radius from centimeters (cm) to meters (m): 200 cm = 2.00 m.

$$a_c = \frac{v^2}{R} = \frac{(4)^2}{2} = \frac{16}{2} = 8.0 \text{ m/s}^2$$

The acceleration is $a_c = 8.0$ m/s^2.

(B) Determine the tension in the tail when the mouse is at the bottom of its arc.

Solution. Draw and label a FBD for the mouse.

- **Weight** ($m\vec{g}$) pulls straight down.
- **Tension** ($\vec{T}$) pulls along the tail, which is straight up when the mouse is at the bottom of its arc.
- The inward (in) direction is toward the center of the circle: straight up when the mouse is at the bottom of its arc.
- The tangential (tan) direction is along a tangent: horizontal for the position shown.

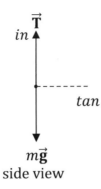

side view

Apply Newton's second law to the mouse. Tension points inward, so it is positive. Weight points outward, so it is negative.

$$\sum F_{in} = ma_c$$
$$T - mg = ma_c$$

Add weight to both sides of the equation.

$$T = mg + ma_c$$

Convert the mass from grams (g) to kilograms (kg).

$$m = 500 \text{ g} = 0.500 \text{ kg} = \frac{1}{2} \text{kg}$$

Plug this value for mass and the acceleration from part (A) into the equation for tension.

$$T = \frac{1}{2}(9.81) + \frac{1}{2}8 = 4.905 + 4 = 8.9 \text{ N}$$

The tension is $T = 8.9$ N.

Example 103. A 50-kg monkey defies gravity on an amusement park centrifuge, which consists of a large cylindrical room, as illustrated below. The monkey stands against the wall. First, the centrifuge rotates with constant angular speed, then the floor disappears! Yet, the monkey does not slide down the wall. The coefficient of friction between the monkey and the wall is $\frac{1}{4}$. The radius of the centrifuge is 10 m.

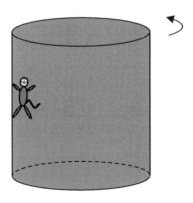

(A) What minimum acceleration does the monkey need in order to not to slide downward?

Solution. Draw and label a FBD for the monkey.

- **Weight** ($m\vec{g}$) pulls straight down.
- **Normal force** ($\vec{N}$) is perpendicular to the wall: it is horizontal (to the right for the position of the monkey shown).
- **Friction** ($\vec{f}_s$) is along the surface: it is up. Friction prevents the monkey from falling.
- The inward (*in*) direction is toward the center of the circle: to the right for the position of the monkey shown.
- The z-direction is perpendicular to the plane of the circle: it is vertical.

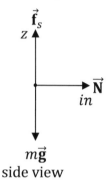

side view

Apply Newton's second law to the monkey. Normal force is directed inward, friction is along $+z$, and weight is along $-z$.

$$\sum F_{in} = ma_c \quad , \quad \sum F_z = 0$$
$$N = ma_c \quad , \quad f_s - mg = 0$$

Solve for the friction force in the inward sum.

$$f_s = mg = 50(9.81) = 491 \text{ N}$$

The force (f_s) of **static** friction is less than or equal to the coefficient (μ_s) of static friction times normal force (N). It's static, not kinetic, because the monkey isn't sliding up or down.

$$f_s \leq \mu_s N$$

Substitute the expression for normal force from the inward sum into the friction inequality.

$$f_s \leq \mu_s m a_c$$

Divide both sides of the inequality by $\mu_s m$.

$$a_c \geq \frac{f_s}{\mu_s m} = \frac{491}{\left(\frac{1}{4}\right)(50)} = 39 \text{ m/s}^2$$

The acceleration is $a_c \geq 39$ m/s^2. Therefore, the minimum acceleration needed is 39 m/s^2.

(B) What minimum speed does the monkey need in order to not to slide downward?

Use the equation for centripetal acceleration.

$$a_c = \frac{v^2}{R}$$

Multiply both sides of the equation by the radius.

$$a_c R = v^2$$

Squareroot both sides of the equation.

$$v = \sqrt{a_c R}$$

$$v \geq \sqrt{(39)(10)} = \sqrt{390} = 20 \text{ m/s}$$

The speed is $v \geq 20$ m/s. Therefore, the minimum speed needed is 20 m/s.

Example 104. A circular racetrack has a diameter of $180\sqrt{3}$ m (the radius R is labeled below) and constant banking angle of $30°$ (as shown below, with the racecar headed out of the page). The racetrack is frictionless. What speed does the racecar need to have in order not to slide up or down the bank?

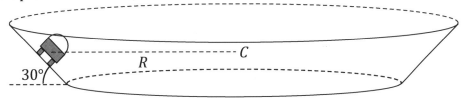

side view

Solution. Draw and label a FBD for the racecar.

- **Weight** ($m\vec{g}$) pulls straight down.
- **Normal force** ($\vec{N}$) is perpendicular to the road (see below).
- The inward (*in*) direction is toward the center of the circle: to the right for the position of the racecar shown.
- The z-direction is perpendicular to the plane of the circle: it is vertical.

side view

Apply Newton's second law to the racecar. Normal force doesn't lie on an axis, so it goes in both sums with trig: The inward component of normal force has a sine function because inward is opposite to $30°$, while the z-component of normal force has a cosine function because the z-axis is adjacent to $30°$. Weight pulls in the $-z$-direction.

$$\sum F_{in} = ma_c \quad , \quad \sum F_z = 0$$

$$N \sin 30° = ma_c \quad , \quad N \cos 30° - mg = 0$$

First add mg to both sides of the second equation.

$$N \sin 30° = ma_c \quad , \quad N \cos 30° = mg$$

Next divide the two equations. Divide the left-hand sides and divide the right-hand sides.

$$\frac{N \sin 30°}{N \cos 30°} = \frac{ma_c}{mg}$$

Normal force and mass cancel out.

$$\frac{\sin 30°}{\cos 30°} = \frac{a_c}{g}$$

Sine divided by cosine makes a tangent: $\tan \theta = \frac{\sin \theta}{\cos \theta}$.

$$\tan 30° = \frac{a_c}{g}$$

Multiply both sides of the equation by gravity.

$$a_c = g \tan 30°$$

$$a_c = (9.81)\frac{\sqrt{3}}{3} = 5.77 \text{ m/s}^2$$

Use the equation for centripetal acceleration.

$$a_c = \frac{v^2}{R}$$

Multiply both sides of the equation by the radius.

$$a_c R = v^2$$

Squareroot both sides of the equation.

$$v = \sqrt{a_c R}$$

Divide the given diameter by two in order to find the radius.

$$R = \frac{D}{2} = \frac{180\sqrt{3}}{2} = 90\sqrt{3} \text{ m}$$

Plug the radius and acceleration into the previous equation for speed.

$$v \geq \sqrt{(5.77)(90\sqrt{3})} = 30 \text{ m/s}$$

The speed is $v = 30$ m/s.

18 NEWTON'S LAW OF GRAVITY

Equations		
$F_g = G\dfrac{m_1 m_2}{R^2}$	$F_g = mg$	$g = G\dfrac{m_p}{R^2}$

Gravitational Constant
$G = 6.67 \times 10^{-11}\ \dfrac{\text{N·m}^2}{\text{kg}^2} \approx \dfrac{20}{3} \times 10^{-11}\ \dfrac{\text{N·m}^2}{\text{kg}^2} = \dfrac{2}{3} \times 10^{-10}\ \dfrac{\text{N·m}^2}{\text{kg}^2}$

Math note:

$$6.67 \approx \frac{20}{3} = \left(\frac{2}{3}\right)(10) = \frac{2}{3} \times 10^1$$

$$6.67 \times 10^{-11} \approx \frac{20}{3} \times 10^{-11} = \left(\frac{2}{3}\right)(10) \times 10^{-11} = \frac{2}{3} \times 10^1 \times 10^{-11} = \frac{2}{3} \times 10^{-10}$$

Symbol	Name	SI Units
F_g	gravitational force	N
m	mass	kg
mg	weight	N
R	separation	m
G	gravitational constant	$\dfrac{\text{N·m}^2}{\text{kg}^2}$ or $\dfrac{\text{m}^3}{\text{kg·s}^2}$
g	gravitational acceleration	m/s^2

Example 105. Planet Ban has a radius of 6.0×10^7 m. Gravitational acceleration near the surface of Ban is 20 m/s².

(A) What is the mass of Ban?

Solution. First identify the information given in the problem.

- The radius of the planet is $R = 6.0 \times 10^7$ m.
- Gravitational acceleration near the surface of the planet is $g = 20$ m/s².
- We will express the gravitational constant as a fraction, $G \approx \frac{2}{3} \times 10^{-10} \frac{\text{N·m}^2}{\text{kg}^2}$.
- We are looking for the mass of the planet, m_p.

Use the equation for gravitational acceleration.

$$g = G\frac{m_p}{R^2}$$

Multiply both sides of the equation by R^2.

$$gR^2 = Gm_p$$

Divide both sides of the equation by G.

$$m_p = \frac{gR^2}{G} = \frac{(20)(6.0 \times 10^7)^2}{\frac{2}{3} \times 10^{-10}}$$

Note that $(6.0 \times 10^7)^2 = (6)^2 \times (10^7)^2$ according to the rule $(ab)^m = a^m b^m$. To divide by the fraction $\frac{2}{3}$, multiply by its **reciprocal**, which is $\frac{3}{2}$.

$$m_p = \frac{(20)(6)^2(10^7)^2}{10^{-10}} \times \frac{3}{2}$$

It's convenient to regroup this to separate the powers of 10. Since multiplication is commutative, the order of multiplication doesn't matter: For example, $20 \times 3 = 3 \times 20$.

$$m_p = \frac{(20)(6)^2(3)}{2} \times \frac{(10^7)^2}{10^{-10}} = 1080 \times \frac{10^{14}}{10^{-10}} = 1080 \times 10^{24} = 1.08 \times 10^{27} \text{ kg}$$

Note that $(10^7)^2 = 10^{14}$ according to the rule $(x^a)^b = x^{ab}$ and that $\frac{10^{14}}{10^{-10}} = 10^{14-(-10)} = 10^{14+10} = 10^{24}$ according to the rule $\frac{x^m}{x^n} = x^{m-n}$. The mass of Ban is $m_p = 1.08 \times 10^{27}$ kg.

(B) Ban has a moon named Ana that has an orbital radius of 4.0×10^9 m and a mass of 2.0×10^{22} kg. What is the force of attraction between Ban and Ana?

Solution. First identify the information given in the problem.

- The separation between the centers of Ban and Ana is $R = 4.0 \times 10^9$ m.
- The mass of Ana is $m_A = 2.0 \times 10^{22}$ kg.
- In part (A), we found that the mass of Ban is $m_p = 1.08 \times 10^{27}$ kg.
- We are looking for gravitational force, F_g.

Use the equation for gravitational force (Newton's law of gravity).

$$F_g = G\frac{m_1 m_2}{R^2}$$

$$F_g = \left(\frac{2}{3} \times 10^{-10}\right) \frac{(1.08 \times 10^{27})(2.0 \times 10^{22})}{(4.0 \times 10^9)^2}$$

Separate the powers of 10. Note that $(4.0 \times 10^9)^2 = (4)^2(10^9)^2$ according to the rule $(ab)^m = a^m b^m$.

$$F_g = \frac{(2)(1.08)(2)}{(3)(4)^2} \frac{10^{-10}10^{27}10^{22}}{(10^9)^2} = \frac{4.32}{48} \frac{10^{39}}{10^{18}} = .090 \times 10^{21} = 9.0 \times 10^{19} \text{ N}$$

Note that $10^{-10}10^{27}10^{22} = 10^{-10+27+22} = 10^{39}$ according to the rule $x^m x^n = x^{m+n}$ and that $\frac{10^{39}}{10^{18}} = 10^{39-18} = 10^{21}$ according to the rule $\frac{x^m}{x^n} = x^{m-n}$. The force is $F_g = 9.0 \times 10^{19}$ N.

Example 106. (A) A planet has 5 times earth's radius and 3 times the surface gravity of earth. How does that planet's mass compare to earth's?

Solution. It's not necessary to look up the mass of the earth or the radius of the earth. Express the given numbers as ratios. The planet's radius is 5 times earth's radius.

$$\frac{R_p}{R_e} = 5$$

The planet's surface gravity is 3 times earth's surface gravity.

$$\frac{g_p}{g_e} = 3$$

Divide the equation for surface gravity on the planet $\left(g_p = \frac{Gm_p}{R_p^2}\right)$ by the equation for the surface gravity on earth $\left(g_e = \frac{Gm_e}{R_e^2}\right)$.

$$\frac{g_p}{g_e} = \frac{\dfrac{Gm_p}{R_p^2}}{\dfrac{Gm_e}{R_e^2}}$$

Divide the two fractions. This means to multiply by the **reciprocal**.

$$\frac{g_p}{g_e} = \frac{Gm_p}{R_p^2} \div \frac{Gm_e}{R_e^2} = \frac{Gm_p}{R_p^2} \times \frac{R_e^2}{Gm_e} = \frac{m_p}{m_e} \frac{R_e^2}{R_p^2}$$

Solve for the ratio $\frac{m_p}{m_e}$. Multiply both sides by R_p^2 and divide both sides by R_e^2.

$$\frac{m_p}{m_e} = \frac{g_p}{g_e} \frac{R_p^2}{R_e^2} = \frac{g_p}{g_e} \left(\frac{R_p}{R_e}\right)^2 = (3)(5)^2 = (3)(25) = 75$$

The answer is $m_p = 75m_e$. The planet's mass is 75× greater than earth's.

(B) A planet has 6 times earth's mass and 96 times the surface gravity of earth. How does that planet's radius compare to earth's?

Solution. Express the given numbers as ratios. The planet's mass is 6 times earth's mass.

$$\frac{m_p}{m_e} = 6$$

The planet's surface gravity is 96 times earth's surface gravity.

$$\frac{g_p}{g_e} = 96$$

Divide the equation for surface gravity on the planet ($g_p = \frac{Gm_p}{R_p^2}$) by the equation for the surface gravity on earth ($g_e = \frac{Gm_e}{R_e^2}$).

$$\frac{g_p}{g_e} = \frac{\dfrac{Gm_p}{R_p^2}}{\dfrac{Gm_e}{R_e^2}}$$

Divide the two fractions. This means to multiply by the **reciprocal**.

$$\frac{g_p}{g_e} = \frac{Gm_p}{R_p^2} \div \frac{Gm_e}{R_e^2} = \frac{Gm_p}{R_p^2} \times \frac{R_e^2}{Gm_e} = \frac{m_p R_e^2}{m_e R_p^2}$$

Solve for the ratio $\frac{R_p}{R_e}$. Multiply both sides by R_p^2 and divide both sides by R_e^2.

$$\frac{g_p}{g_e} \frac{R_p^2}{R_e^2} = \frac{m_p}{m_e}$$

Multiply both sides by g_e and divide both sides by g_p.

$$\frac{R_p^2}{R_e^2} = \frac{m_p}{m_e} \frac{g_e}{g_p}$$

Squareroot both sides of the equation.

$$\frac{R_p}{R_e} = \sqrt{\frac{m_p}{m_e} \frac{g_e}{g_p}} = \sqrt{(6)\left(\frac{1}{96}\right)} = \sqrt{\frac{1}{16}} = \frac{1}{4}$$

The answer is $R_p = \frac{R_e}{4}$. The planet's radius is $\frac{1}{4}\times$ earth's radius.

Example 107. Planet Mon has a mass of 1.6×10^{25} kg. Planet Mon has a moon named Key with a mass of 2.5×10^{23} kg. The distance between the center of Mon and Key is 9.0×10^8 m. Find the point where the net gravitational field equals zero.
Solution. Begin with a labeled diagram.

Note that, in this example, R is **not** the radius of a planet or moon: R is the distance from the center of the planet or moon to the point where we're calculating g. The only time that R is the radius of a planet is when you're trying to find g at its surface.

In between the planet and moon, the two gravitational fields will be in opposite directions. At some point in between, the two gravitational fields will also have equal magnitude and therefore cancel out. To figure out where this special point is, set the magnitudes of the two gravitational fields equal to one another ($g_m = g_k$).

$$G\frac{m_m}{R_m^2} = G\frac{m_k}{R_k^2}$$

Divide both sides of the equation by the gravitational constant: G cancels out.

$$\frac{m_m}{R_m^2} = \frac{m_k}{R_k^2}$$

Cross-multiply in order to remove the unknowns from the denominator:

$$R_k^2 m_m = R_m^2 m_k$$

We have one equation, but two unknowns: The unknowns are R_m and R_k. This means that we need a second equation. From the picture on the previous page, we can see that R_m and R_k add up to the distance between the centers of Mon and Key, d, where $d = 9.0 \times 10^8$ m.

$$R_m + R_k = d$$

Isolate one of the unknowns in this equation:

$$R_k = d - R_m$$

Substitute this expression in place of R_k in the equation $R_k^2 m_m = R_m^2 m_k$:

$$(d - R_m)^2 m_m = R_m^2 m_k$$

If you foil the left-hand side and distribute, you get a quadratic equation for R_m. You can't always avoid a quadratic, but in this case you can avoid the quadratic by taking the square-root of both sides.

$$\sqrt{(d - R_m)^2 m_m} = \sqrt{R_m^2 m_k}$$

Note that $\sqrt{(d - R_m)^2} = \pm(d - R_m)$ and $\sqrt{R_m^2} = \pm R_m$. The $\pm$ is there because $\sqrt{x^2} = \pm x$. For example, $\sqrt{4}$ has two solutions, -2 and $+2$, since $(-2)^2 = 4$ and $(+2)^2 = 4$. We must consider both signs when solving for R_m. Note that $\sqrt{(d - R_m)^2 m_m} = \pm(d - R_m)\sqrt{m_m}$ and $\sqrt{R_m^2 m_k} = \pm R_m\sqrt{m_k}$.

$$(d - R_m)\sqrt{m_m} = \pm R_m\sqrt{m_k}$$

Distribute the $\sqrt{m_m}$ to both the d and $-R_m$.

$$d\sqrt{m_m} - R_m\sqrt{m_m} = \pm R_m\sqrt{m_k}$$

Add $R_m\sqrt{m_m}$ to both sides in order to group the unknown (R_m) terms together.

$$d\sqrt{m_m} = R_m\sqrt{m_m} \pm R_m\sqrt{m_k}$$

Factor out the R_m.

$$d\sqrt{m_m} = R_m(\sqrt{m_m} \pm \sqrt{m_k})$$

Divide both sides of the equation by $\sqrt{m_m} \pm \sqrt{m_k}$.

$$R_m = \frac{d\sqrt{m_m}}{\sqrt{m_m} \pm \sqrt{m_k}}$$

Only the positive root yields an answer which lies in between the two masses.

$$R_m = \frac{(9.0 \times 10^8)\sqrt{1.6 \times 10^{25}}}{\sqrt{1.6 \times 10^{25}} + \sqrt{2.5 \times 10^{23}}} = \frac{(9 \times 10^8)(4 \times 10^{13})}{4 \times 10^{13} + 5 \times 10^{12}} = \frac{36 \times 10^{21}}{4.5 \times 10^{13}} = 8.0 \times 10^8 \text{ m}$$

The answer is $R_m = 8.0 \times 10^8$ m from Planet Mon (or $R_k = 1.0 \times 10^8$ m from Key).

Example 108. Planet Coco has a mass of 3.0×10^{24} kg. Planet Nut has a mass of 6.0×10^{24} kg. The distance between the center of Coco and Nut is 4.0×10^8 m. What are the magnitude and direction of the net gravitational field at the point exactly halfway between the centers of these two planets?

Solution. Begin with a labeled diagram.

First calculate the magnitude of each individual gravitational field at the specified point. Note that R_c and R_n are **not** the radii of the two planets because we're not finding gravitational acceleration on their surfaces. We're finding the gravitational field at the point **halfway** between the two planets. Therefore, $R_c = R_n = \frac{d}{2} = \frac{4.0 \times 10^8}{2} = 2.0 \times 10^8$ m.

$$g_c = G\frac{m_c}{R_c^2} = \left(\frac{2}{3} \times 10^{-10}\right)\frac{(3.0 \times 10^{24})}{(2 \times 10^8)^2} = \frac{(2)(3)}{(3)(2)^2} \times \frac{10^{-10}10^{24}}{(10^8)^2} = \frac{6}{12} \times 10^{-2} = \frac{1}{200} \text{ m/s}^2$$

$$g_n = G\frac{m_n}{R_n^2} = \left(\frac{2}{3} \times 10^{-10}\right)\frac{(6.0 \times 10^{24})}{(2 \times 10^8)^2} = \frac{(2)(6)}{(3)(2)^2} \times \frac{10^{-10}10^{24}}{(10^8)^2} = \frac{12}{12} \times 10^{-2} = \frac{1}{100} \text{ m/s}^2$$

Midway between the two planets, the gravitational fields will point in opposite directions. Therefore, we must subtract the two gravitational fields to find the net gravitational field at the midpoint.

$$g_{net} = \frac{1}{g_n} - \frac{1}{g_c} = \frac{1}{100} - \frac{1}{200} = \frac{2}{200} - \frac{1}{200} = \frac{2-1}{200} = \frac{1}{200} \text{ m/s}^2$$

The answer is $g_{net} = \frac{1}{200}$ m/s^2, which is the same as $g_{net} = 0.0050$ m/s^2.

19 SATELLITE MOTION

Equations	
acceleration	$a_c = G\dfrac{m_p}{R^2}$, $a_c = \dfrac{v^2}{R}$
speed	$v = \sqrt{G\dfrac{m_p}{R}}$, $v = \dfrac{2\pi R}{T}$
period	$T = 2\pi\sqrt{\dfrac{R^3}{Gm_p}}$, $T = \dfrac{2\pi R}{v}$
altitude	$h = R - R_p$

Symbol	Name	SI Units
m_p	mass of the large astronomical body near the center of the orbit	kg
m_s	mass of the satellite	kg
R	radius of the satellite's orbit	m
R_p	radius of the planet	m
h	altitude	m
G	gravitational constant	$\dfrac{\text{N·m}^2}{\text{kg}^2}$ or $\dfrac{\text{m}^3}{\text{kg·s}^2}$
a_c	centripetal acceleration	m/s^2
v	speed of the satellite	m/s
T	period of the satellite's revolution	s

Example 109. Planet Φ_6 has a mass of 1.2×10^{24} kg. A satellite orbits Φ_6 in a circular orbit with a radius of 2.0×10^7 m.

(A) What is the satellite's orbital speed?

Solution. First identify the information given in the problem.

- The radius of the satellite's orbit is $R = 2.0 \times 10^7$ m.
- The mass of the planet is $m_p = 1.2 \times 10^{24}$ kg.
- We will express the gravitational constant as a fraction, $G \approx \frac{2}{3} \times 10^{-10} \frac{\text{N·m}^2}{\text{kg}^2}$.
- We are looking for the speed of the satellite, v.

Use the equation for satellite speed that involves G, m_p, and R.

$$v = \sqrt{G \frac{m_p}{R}} = \sqrt{\left(\frac{2}{3} \times 10^{-10}\right) \frac{(1.2 \times 10^{24})}{(2.0 \times 10^7)}}$$

It's convenient to regroup this to separate the powers of 10.

$$v = \sqrt{\frac{(2)(1.2)}{(3)(2)} \frac{10^{-10} 10^{24}}{10^7}} = \sqrt{\frac{2.4}{6} \frac{10^{14}}{10^7}} = \sqrt{0.4 \times 10^7} = \sqrt{4 \times 10^6}$$

Note that $10^{-10} 10^{24} = 10^{-10+24} = 10^{14}$ according to the rule $(x^a)^b = x^{ab}$ and that $\frac{10^{14}}{10^7} = 10^{14-7} = 10^7$ according to the rule $\frac{x^m}{x^n} = x^{m-n}$. Also note that $0.4 \times 10^7 = 4 \times 10^6$.

$$v = \sqrt{4}\sqrt{10^6} = 2.0 \times 10^3 \text{ m/s} = 2000 \text{ m/s} = 2.0 \text{ km/s}$$

Note that $2.0 \times 10^3 = 2000$ and that 2000 meters per second is the same as 2 kilometers per second since the metric prefix kilo (k) stands for one thousand (1000). The orbital speed of the satellite is $v = 2.0$ km/s.

(B) What is the satellite's orbital period?

Solution. Since we already found the speed in part (A), we can use the equation for period that involves R and v.

$$v = \frac{2\pi R}{T}$$

Multiply both sides of the equation by T.

$$vT = 2\pi R$$

Divide both sides of the equation by v.

$$T = \frac{2\pi R}{v} = \frac{2\pi(2.0 \times 10^7)}{(2.0 \times 10^3)} = 2\pi \times 10^4 \text{ s} = 6.3 \times 10^4 \text{ s}$$

The orbital period is $T = 2\pi \times 10^4$ s, which is the same as 6.3×10^4 s.

Example 110. A satellite orbits the earth in a geosynchronous circular orbit above the equator. Approximate earth's mass as $\approx 6.0 \times 10^{24}$ kg. What is the radius of the satellite's orbit?

Solution. First identify the information given in the problem.

- The mass of the planet is $m_p \approx 6.0 \times 10^{24}$ kg.

- This satellite is **geosynchronous**. What this means is that the satellite's orbit is synchronized with earth's rotation. It takes the earth 24 hours to complete one rotation on its axis, so a geosynchronous satellite also has a period of 24 hours. We must convert this to seconds: $24 \text{ hr} = 24 \times 60 \text{ min.} = 24 \times 60 \times 60 \text{ s} = 86{,}400 \text{ s}$. The period is 86,400 s.

- We will express the gravitational constant as a fraction, $G \approx \frac{2}{3} \times 10^{-10} \; \frac{\text{N·m}^2}{\text{kg}^2}$.

- We are looking for the radius of the satellite's orbit, R.

Use the equation for period that involves G, m_p, and R.

$$T = 2\pi \sqrt{\frac{R^3}{Gm_p}}$$

Divide both sides of the equation by 2π.

$$\frac{T}{2\pi} = \sqrt{\frac{R^3}{Gm_p}}$$

Square both sides of the equation.

$$\frac{T^2}{4\pi^2} = \frac{R^3}{Gm_p}$$

Multiply both sides of the equation by Gm_p.

$$R^3 = \frac{Gm_p T^2}{4\pi^2}$$

Take the cube root of both sides of the equation.

$$R = \sqrt[3]{\frac{Gm_p T^2}{4\pi^2}} = \sqrt[3]{\frac{\left(\frac{2}{3} \times 10^{-10}\right)(6.0 \times 10^{24})(86{,}400)^2}{4\pi^2}} = \sqrt[3]{\frac{746496}{\pi^2} \times 10^{18}} = 4.23 \times 10^7 \text{ m}$$

The orbital radius is $R = 4.23 \times 10^7$ m (which can also be expressed as $72 \times \sqrt[3]{\frac{2}{\pi^2}} \times 10^6$ m.)

Note that one way to perform a cube root on a calculator is to raise the quantity to the power of (1/3) using parentheses. For example, write (2/3.14^2)^(1/3) to evaluate $\sqrt[3]{\frac{2}{\pi^2}}$ on a calcuator.

Where Do the Satellite Equations Come From?

Consider a satellite traveling with constant speed in a circle. The net force is a gravitational force directed inward, which causes centripetal acceleration. Apply Newton's second law to uniform circular motion as described in Chapter 17.

$$\sum F_{in} = m_s a_c$$

Newton's law of gravity provides the equation $\left(F_g = G \frac{m_p m_s}{R^2} \right)$ for the force:

$$G \frac{m_p m_s}{R^2} = m_s a_c$$

The satellite's mass appears on the right-hand side because it is the satellite which is traveling in a large circle. Divide both sides by the mass of the satellite to solve for the **acceleration** (a_c) of the satellite:

$$a_c = G \frac{m_p}{R^2}$$

Use the equation for centripetal acceleration $\left(a_c = \frac{v^2}{R} \right)$ from Chapter 16:

$$\frac{v^2}{R} = G \frac{m_p}{R^2}$$

Multiply both sides by R to solve for the **speed** (v) of the satellite:

$$v^2 = G \frac{m_p}{R} \quad \text{or} \quad v = \sqrt{G \frac{m_p}{R}}$$

To find the period of the satellite, use an equation from Chapter 16 $\left(v = \frac{2\pi R}{T} \right)$:

$$\left(\frac{2\pi R}{T} \right)^2 = G \frac{m_p}{R} \quad \text{or} \quad \frac{4\pi^2 R^2}{T^2} = G \frac{m_p}{R}$$

Cross-multiply to solve for the **period** (T):

$$4\pi^2 R^3 = G m_p T^2$$

$$T^2 = \frac{4\pi^2 R^3}{G m_p} \quad \text{or} \quad T = 2\pi \sqrt{\frac{R^3}{G m_p}}$$

Note that R is the radius of the satellite's orbit. It is **not** the radius of the planet.

20 THE SCALAR PRODUCT

Equations	
$\vec{A} \cdot \vec{B} = A_x B_x + A_y B_y + A_z B_z$	$\vec{A} \cdot \vec{B} = AB \cos\theta$
$\vec{A} = A_x \hat{x} + A_y \hat{y} + A_z \hat{z}$	$\vec{B} = B_x \hat{x} + B_y \hat{y} + B_z \hat{z}$
$A = \sqrt{A_x^2 + A_y^2 + A_z^2}$	$B = \sqrt{B_x^2 + B_y^2 + B_z^2}$

Scalar Product between Unit Vectors		
$\hat{x} \cdot \hat{x} = 1$	$\hat{y} \cdot \hat{y} = 1$	$\hat{z} \cdot \hat{z} = 1$
$\hat{x} \cdot \hat{y} = 0$	$\hat{y} \cdot \hat{z} = 0$	$\hat{z} \cdot \hat{x} = 0$
$\hat{y} \cdot \hat{x} = 0$	$\hat{z} \cdot \hat{y} = 0$	$\hat{x} \cdot \hat{z} = 0$

Example 111. Find the scalar product between the following vectors.

$$\vec{A} = 6\,\hat{x} + \hat{y} + 3\,\hat{z}$$
$$\vec{B} = 2\,\hat{x} + 8\,\hat{y} + 4\,\hat{z}$$

Solution. First identify the components of the given vectors.

- Compare $\vec{A} = 6\,\hat{x} + \hat{y} + 3\,\hat{z}$ to $\vec{A} = A_x \hat{x} + A_y \hat{y} + A_z \hat{z}$. The components of vector $\vec{A}$ are $A_x = 6$, $A_y = 1$, and $A_z = 3$.
- Compare $\vec{B} = 2\,\hat{x} + 8\,\hat{y} + 4\,\hat{z}$ to $\vec{B} = B_x \hat{x} + B_y \hat{y} + B_z \hat{z}$. The components of vector $\vec{B}$ are $B_x = 2$, $B_y = 8$, and $B_z = 4$.

Use the component form of the scalar product.

$$\vec{A} \cdot \vec{B} = A_x B_x + A_y B_y + A_z B_z = (6)(2) + (1)(8) + (3)(4) = 12 + 8 + 12 = 32$$

The scalar product is $\vec{A} \cdot \vec{B} = 32$.

Example 112. Find the scalar product between the following vectors.

$$\vec{A} = 2\,\hat{x} - \hat{y} + 6\,\hat{z}$$
$$\vec{B} = -3\,\hat{x} + 2\,\hat{y} + 5\,\hat{z}$$

Solution. First identify the components of the given vectors.

- Compare $\vec{A} = 2\,\hat{x} - \hat{y} + 6\,\hat{z}$ to $\vec{A} = A_x\hat{x} + A_y\hat{y} + A_z\hat{z}$. The components of vector $\vec{A}$ are $A_x = 2, A_y = -1$, and $A_z = 6$.
- Compare $\vec{B} = -3\,\hat{x} + 2\,\hat{y} + 5\,\hat{z}$ to $\vec{B} = B_x\hat{x} + B_y\hat{y} + B_z\hat{z}$. The components of vector $\vec{B}$ are $B_x = -3, B_y = 2$, and $B_z = 5$.

Use the component form of the scalar product.

$$\vec{A} \cdot \vec{B} = A_xB_x + A_yB_y + A_zB_z = (2)(-3) + (-1)(2) + (6)(5) = -6 - 2 + 30 = 22$$

The scalar product is $\vec{A} \cdot \vec{B} = 22$.

Example 113. Find the scalar product between the following vectors.

$$\vec{A} = 3\,\hat{y} + 4\,\hat{z}$$
$$\vec{B} = 9\,\hat{x} - 3\,\hat{y} - \hat{z}$$

Solution. First identify the components of the given vectors.

- Compare $\vec{A} = 3\,\hat{y} + 4\,\hat{z}$ to $\vec{A} = A_x\hat{x} + A_y\hat{y} + A_z\hat{z}$. The components of vector $\vec{A}$ are $A_x = 0, A_y = 3$, and $A_z = 4$.
- Compare $\vec{B} = 9\,\hat{x} - 3\,\hat{y} - \hat{z}$ to $\vec{B} = B_x\hat{x} + B_y\hat{y} + B_z\hat{z}$. The components of vector $\vec{B}$ are $B_x = 9, B_y = -3$, and $B_z = -1$.

Use the component form of the scalar product.

$$\vec{A} \cdot \vec{B} = A_xB_x + A_yB_y + A_zB_z = (0)(9) + (3)(-3) + (4)(-1) = 0 - 9 - 4 = -13$$

The scalar product is $\vec{A} \cdot \vec{B} = -13$.

Example 114. Find the scalar product between the following vectors.

$$\vec{A} = 12\,\hat{x} + 4\,\hat{y} - 16\,\hat{z}$$
$$\vec{B} = 2\,\hat{x} - 24\,\hat{y} - 6\,\hat{z}$$

Solution. First identify the components of the given vectors.

- Compare $\vec{A} = 12\,\hat{x} + 4\,\hat{y} - 16\,\hat{z}$ to $\vec{A} = A_x\hat{x} + A_y\hat{y} + A_z\hat{z}$. The components of vector $\vec{A}$ are $A_x = 12, A_y = 4$, and $A_z = -16$.
- Compare $\vec{B} = 2\,\hat{x} - 24\,\hat{y} - 6\,\hat{z}$ to $\vec{B} = B_x\hat{x} + B_y\hat{y} + B_z\hat{z}$. The components of vector $\vec{B}$ are $B_x = 2, B_y = -24$, and $B_z = -6$.

Use the component form of the scalar product.

$$\vec{A} \cdot \vec{B} = A_xB_x + A_yB_y + A_zB_z = (12)(2) + (4)(-24) + (-16)(-6) = 24 - 96 + 96 = 24$$

The scalar product is $\vec{A} \cdot \vec{B} = 24$.

Example 115. Determine the angle between the following vectors.

$$\vec{A} = 3\,\hat{y} - 3\,\hat{z}$$
$$\vec{B} = -6\,\hat{y} + 6\,\hat{z}$$

Solution. First identify the components of the given vectors.

- Compare $\vec{A} = 3\,\hat{y} - 3\,\hat{z}$ to $\vec{A} = A_x\hat{x} + A_y\hat{y} + A_z\hat{z}$. The components of vector $\vec{A}$ are $A_x = 0$, $A_y = 3$, and $A_z = -3$.
- Compare $\vec{B} = -6\,\hat{y} + 6\,\hat{z}$ to $\vec{B} = B_x\hat{x} + B_y\hat{y} + B_z\hat{z}$. The components of vector $\vec{B}$ are $B_x = 0$, $B_y = -6$, and $B_z = 6$.

Use the component form of the scalar product.

$$\vec{A}\cdot\vec{B} = A_xB_x + A_yB_y + A_zB_z = (0)(0) + (3)(-6) + (-3)(6) = 0 - 18 - 18 = -36$$

Next find the magnitude and direction of the given vectors.

$$A = \sqrt{A_x^2 + A_y^2 + A_z^2} = \sqrt{(0)^2 + (3)^2 + (-3)^2} = \sqrt{0 + 9 + 9} = \sqrt{18} = \sqrt{(9)(2)} = 3\sqrt{2}$$

$$B = \sqrt{B_x^2 + B_y^2 + B_z^2} = \sqrt{(0)^2 + (-6)^2 + (6)^2} = \sqrt{0 + 36 + 36} = \sqrt{72} = \sqrt{(36)(2)} = 6\sqrt{2}$$

Set the two forms of the scalar product equal to each other.

$$A_xB_x + A_yB_y + A_zB_z = AB\cos\theta$$

We already found that the scalar product is $A_xB_x + A_yB_y + A_zB_z = -36$, the magnitude of $\vec{A}$ is $A = 3\sqrt{2}$, and the magnitude of $\vec{B}$ is $B = 6\sqrt{2}$. Plug these values into the previous equation.

$$-36 = \left(3\sqrt{2}\right)\left(6\sqrt{2}\right)\cos\theta$$
$$-36 = (3)(6)\sqrt{2}\sqrt{2}\cos\theta$$
$$-36 = 18\sqrt{2}\sqrt{2}\cos\theta$$

Note that $\sqrt{2}\sqrt{2} = 2$.

$$-36 = 18(2)\cos\theta$$
$$-36 = 36\cos\theta$$

Divide both sides of the equation by 36.

$$\cos\theta = -\frac{36}{36}$$
$$\cos\theta = -1$$
$$\theta = \cos^{-1}(-1)$$
$$\theta = 180°$$

The angle between $\vec{A}$ and $\vec{B}$ is $180°$. Recall that $\cos(180°) = -1$.

Example 116. Determine the angle between the following vectors.

$$\vec{A} = 3\,\hat{x} - \sqrt{3}\,\hat{y} + 2\,\hat{z}$$
$$\vec{B} = \sqrt{3}\,\hat{x} - \hat{y} + 2\sqrt{3}\,\hat{z}$$

Solution. First identify the components of the given vectors.

- Compare $\vec{A} = 3\,\hat{x} - \sqrt{3}\,\hat{y} + 2\,\hat{z}$ to $\vec{A} = A_x\hat{x} + A_y\hat{y} + A_z\hat{z}$. The components of vector $\vec{A}$ are $A_x = 3$, $A_y = -\sqrt{3}$, and $A_z = 2$.
- Compare $\vec{B} = \sqrt{3}\,\hat{x} - \hat{y} + 2\sqrt{3}\,\hat{z}$ to $\vec{B} = B_x\hat{x} + B_y\hat{y} + B_z\hat{z}$. The components of vector $\vec{B}$ are $B_x = \sqrt{3}$, $B_y = -1$, and $B_z = 2\sqrt{3}$.

Use the component form of the scalar product.

$$\vec{A} \cdot \vec{B} = A_xB_x + A_yB_y + A_zB_z = (3)(\sqrt{3}) + (-\sqrt{3})(-1) + (2)(2\sqrt{3}) = 3\sqrt{3} + \sqrt{3} + 4\sqrt{3}$$
$$\vec{A} \cdot \vec{B} = 8\sqrt{3}$$

Next find the magnitude and direction of the given vectors.

$$A = \sqrt{A_x^2 + A_y^2 + A_z^2} = \sqrt{(3)^2 + \left(-\sqrt{3}\right)^2 + (2)^2} = \sqrt{9 + 3 + 4} = \sqrt{16} = 4$$

Note that $\sqrt{3}\sqrt{3} = 3$, $\left(-\sqrt{3}\right)\left(-\sqrt{3}\right) = +3$, and $\left(2\sqrt{3}\right)\left(2\sqrt{3}\right) = 4(3) = 12$.

$$B = \sqrt{B_x^2 + B_y^2 + B_z^2} = \sqrt{\left(\sqrt{3}\right)^2 + (-1)^2 + \left(2\sqrt{3}\right)^2} = \sqrt{3 + 1 + 12} = \sqrt{16} = 4$$

Set the two forms of the scalar product equal to each other.

$$A_xB_x + A_yB_y + A_zB_z = AB\cos\theta$$

We already found that the scalar product is $A_xB_x + A_yB_y + A_zB_z = 8\sqrt{3}$, the magnitude of $\vec{A}$ is $A = 4$, and the magnitude of $\vec{B}$ is $B = 4$. Plug these values into the previous equation.

$$8\sqrt{3} = (4)(4)\cos\theta$$
$$8\sqrt{3} = 16\cos\theta$$

Divide both sides of the equation by 16.

$$\cos\theta = \frac{8\sqrt{3}}{16}$$

Note that $\frac{8}{16} = \frac{1}{2}$. Therefore, $\frac{8\sqrt{3}}{16} = \frac{\sqrt{3}}{2}$.

$$\cos\theta = \frac{\sqrt{3}}{2}$$
$$\theta = \cos^{-1}\left(\frac{\sqrt{3}}{2}\right)$$
$$\theta = 30°$$

The angle between $\vec{A}$ and $\vec{B}$ is $30°$. Recall that $\cos(30°) = \frac{\sqrt{3}}{2}$.

21 WORK AND POWER

Symbol	Name	Units
$\vec{\mathbf{F}}$	force	N
$\vec{\mathbf{s}}$	displacement	m
$d\vec{\mathbf{s}}$	differential displacement	m
θ	angle between $\vec{\mathbf{F}}$ and $\vec{\mathbf{s}}$	° or rad
W	work	J
m	mass	kg
g	gravitational acceleration	m/s^2
Δh	change in height	m
G	gravitational constant	$\frac{N \cdot m^2}{kg^2}$ or $\frac{m^3}{kg \cdot s^2}$
m_p	mass of large astronomical body	kg
R	distance from the center of the planet	m
μ	coefficient of friction	unitless
N	normal force	N
k	spring constant	N/m or kg/s^2
x	displacement of a spring from equilibrium	m
a_x	x-component of acceleration	m/s^2
$\vec{\overline{v}}$	average velocity	m/s
P	instantaneous power	W
$\overline{P}$	average power	W
t	time	s

Work Done by a Variable Force

$$W = \int_i^f \vec{\mathbf{F}} \cdot d\vec{\mathbf{s}} = \int_{x=x_0}^{x} F_x\,dx + \int_{y=y_0}^{y} F_y\,dy + \int_{z=z_0}^{z} F_z\,dz$$

Work Done by Gravity (small change in altitude)

$$W_g = -mg\Delta h$$

Work Done by Gravity (astronomical change in altitude)

$$W_g = Gm_p m \left(\frac{1}{R} - \frac{1}{R_0} \right)$$

Work Done by a Spring

$$W_s = \pm \frac{1}{2} k x^2$$

Work Done by Friction

$$W_{nc} = -\mu N s$$

Work Done by a Constant Force

$$W = \vec{\mathbf{F}} \cdot \vec{\mathbf{s}} = F_x \Delta x + F_y \Delta y + F_z \Delta z = F s \cos\theta$$

Work Done by Normal Force

$$W_N = 0 \text{ (because } \cos 90° = 0)$$

Net Work (where x is along the acceleration)

$$W_{net} = \left(\sum F_x \right) s = m a_x s$$

Instantaneous Power

$$P = \frac{dW}{dt}$$

Average Power

$$\bar{P} = \frac{W}{t}$$

Average Power by a Constant Force

$$\bar{P} = \vec{\mathbf{F}} \cdot \vec{\bar{v}} = F_x \bar{v}_x + F_y \bar{v}_y + F_z \bar{v}_z = F \, \bar{v} \cos\theta$$

Example 117. A monkey drops a 50-kg box of bananas. The box falls 1.5 m. How much work is done by gravity?

Solution. Since this is not an astronomical change in altitude, use the equation $W_g = -mg\Delta h$ to find the work done **by gravity**. Since the box falls, Δh is negative: $\Delta h = -1.5$ m.

$$W_g = -mg\Delta h = -(50)(9.81)(-1.5) = 736 \, \text{J}$$

Two minus signs make the answer positive. The work done by gravity is positive when the final position is below the initial position. The work done by gravity is $W_g = 736$ J.

Example 118. A monkey stretches a spring 4.0 m from equilibrium. The spring constant is 8.0 N/m. How much work is done by the spring?

Solution. Use the equation $W_s = \pm\frac{1}{2}kx^2$ to find the work done **by the spring**.

$$W_s = \pm\frac{1}{2}kx^2$$

The work done "by the spring" is negative because the system travels **away** from the equilibrium position. For this problem, we choose the minus sign.

$$W_s = -\frac{1}{2}kx^2 = -\frac{1}{2}(8)(4)^2 = -4(16) = -64 \, \text{J}$$

The work done by the spring is $W_s = -64$ J.

Example 119. Planet Coconut has a mass of 5.0×10^{24} kg and a radius of 4.0×10^6 m. Find the work done by gravity when a 300-kg rocket climbs from Coconut's surface up to an altitude of 4.0×10^6 m above Coconut's surface.

Solution. For an **astronomical** change in altitude use the equation $W_g = Gm_p m \left(\frac{1}{R} - \frac{1}{R_0}\right)$ to find the work done **by gravity**.

$$W_g = Gm_p m \left(\frac{1}{R} - \frac{1}{R_0}\right)$$

Since the rocket begins on the planet's surface, $R_0 = 4.0 \times 10^6$ m. Use the altitude equation (from Chapter 19) to find R from the given altitude ($h = 4.0 \times 10^6$ m).

$$R = R_0 + h = 4.0 \times 10^6 + 4.0 \times 10^6 = 8.0 \times 10^6 \text{ m}$$

Plug numbers into the previous equation.

$$W_g = \left(\frac{2}{3} \times 10^{-10}\right)(5.0 \times 10^{24})(300)\left(\frac{1}{8.0 \times 10^6} - \frac{1}{4.0 \times 10^6}\right)$$

It is convenient to reorganize and factor out the 10^6.

$$W_g = \frac{(2)(5)(300)}{(3)} \times \frac{10^{-10}10^{24}}{10^6}\left(\frac{1}{8} - \frac{1}{4}\right)$$

Note that $10^{-10}10^{24} = 10^{14}$ according to $x^m x^n = x^{m+n}$ and that $\frac{10^{14}}{10^6} = 10^8$ according to $\frac{x^m}{x^n} = x^{m-n}$. Subtract fractions with a **common denominator**: $\frac{1}{8} - \frac{1}{4} = \frac{1}{8} - \frac{2}{8} = -\frac{1}{8}$.

$$W_g = 1000 \times \frac{10^{14}}{10^6}\left(-\frac{1}{8}\right) = -125 \times 10^8 = -1.25 \times 10^{10} \text{ J}$$

The work done by gravity is $W_g = -1.25 \times 10^{10}$ J. The work done by gravity is negative because the rocket traveled away from the planet.

Example 120. A 200-kg bananamobile uniformly accelerates from rest to 60 m/s in 5.0 s. Find the net work done.

Solution. Use the equation $W_{net} = ma_x \Delta x$ to find the **net work**. Use the equations of **uniform acceleration** (Chapter 3) to find a_x and Δx. The three known quantities for the equations of uniform acceleration are $v_{x0} = 0$, $v_x = 60$ m/s, and $t = 5.0$ s.

$$v_x = v_{x0} + a_x t$$
$$60 = 0 + a_x 5$$
$$a_x = \frac{60}{5} = 12 \text{ m/s}^2$$

Next use an equation of uniform acceleration to find the net displacement.

$$\Delta x = v_{x0}t + \frac{1}{2}a_x t^2 = 0 + \frac{1}{2}(12)(5)^2 = 6(25) = 150 \text{ m}$$

Plug numbers into the equation for the net work.

$$W_{net} = ma_x \Delta x = (200)(12)(150) = 360{,}000 \text{ J}$$

The net work is $W_{net} = 360$ kJ. It's the same as 360,000 J. Recall that the metric prefix kilo (k) stands for 1000.

Example 121. A 20-kg monkey slides 8.0 m down a 30° incline. The coefficient of friction between the monkey and the incline is $\frac{\sqrt{3}}{4}$.

(A) Find the work done by gravity.

Solution. Use the equation $W_g = -mg\Delta h$ to find the work done **by gravity**. The distance given in the problem is $s = 8.0$ m. Draw a right triangle to relate s to Δh.

$$s \quad \Delta h \quad 30°$$

Since the box travels **down** the incline, Δh is negative.

$$\sin 30° = \frac{-\Delta h}{s}$$

$$\Delta h = -4.0 \text{ m}$$

Plug values into the equation for the work done by gravity.

$$W_g = -mg\Delta h = -(20)(9.81)(-4) = 785 \text{ J}$$

Two minus signs make the answer positive. The work done by gravity is positive when the final position is below the initial position. The work done by gravity is $W_g = 785$ J.

(B) Find the work done by the friction force.

Solution. Use the equation $W_{nc} = -\mu N s$ to find the work done **by friction**. (The subscripts nc represent that the work done by friction is nonconservative.) The coefficient of friction is $\mu = \frac{\sqrt{3}}{4}$. Draw a FBD to find **normal force**. (Normal force does **not** equal weight.)

Sum the y-components of the forces to solve for normal force.

$$\sum F_y = ma_y$$

$$N - mg\cos 30° = 0$$

Note that $a_y = 0$ because the box doesn't accelerate perpendicular to the incline.

$$N = mg\cos 30° = (20)(9.81)\frac{\sqrt{3}}{2} = 170 \text{ N}$$

Normal force is $N = 170$ N (it's $N \approx 100\sqrt{3}$ N if you round gravity to 10 m/s²). Plug numbers into the equation for nonconservative work.

$$W_{nc} = -\mu N s = -\frac{\sqrt{3}}{4}(170)(8) = -589 \text{ J}$$

The work done by friction is $W_{nc} = -589$ J. The work is negative because friction subtracts energy from the system.

(C) Find the work done by the normal force.

Solution. Use the equation $W = F s \cos\theta$ to find the work done by the **normal force**. When finding the work done by normal force, $\theta = 90°$ because normal force is perpendicular to the surface. Since $\cos 90° = 0$, normal force doesn't do any work: $W_N = 0$.

(D) Find the net work.

Solution. There are a couple of ways to solve this problem. One method is to use the equation $W_{net} = (\sum F_x)s$ to find the **net work**. Sum the x-components of the forces.

$$\sum F_x = mg \sin 30° - f$$

We must find the friction force before we can calculate the net force ($\sum F_x$). Apply the friction equation.

$$f = \mu N = \frac{\sqrt{3}}{4}(170) = 73.6 \text{ N}$$

Plug values into the previous equation for the net force.

$$\sum F_x = mg \sin 30° - f = (20)(9.81)\frac{1}{2} - 73.6 = 98.1 - 73.6 = 24.5 \text{ N}$$

Use the net force to calculate the net work.

$$W_{net} = \left(\sum F_x\right)s = (24.5)(8) = 196 \text{ J}$$

There was actually a simpler way to solve this problem. Another way to find net work is to find the work done by the individual forces and add them together.

$$W_{net} = W_g + W_{nc} + W_N = 785 - 589 + 0 = 196 \text{ J}$$

Example 122. A force $\vec{F} = 2\,\hat{x} + 6\,\hat{y}$ causes the displacement $\vec{s} = 3\,\hat{x} - 8\,\hat{y}$, where SI units have been suppressed. Find the work done.

Solution. For a constant force (there are no variables in $2\,\hat{x} + 6\,\hat{y}$, since $\hat{x}$ are $\hat{y}$ unit vectors), work is the **scalar product** (Chapter 20) between force and displacement. Identify the components of $\vec{F}$ and $\vec{s}$.

- Compare $\vec{F} = 2\,\hat{x} + 6\,\hat{y}$ to $\vec{F} = F_x\hat{x} + F_y\hat{y} + F_z\hat{z}$. The components of the force vector $\vec{F}$ are $F_x = 2$, $F_y = 6$, and $F_z = 0$.
- Compare $\vec{s} = 3\,\hat{x} - 8\,\hat{y}$ to $\vec{s} = \Delta x\,\hat{x} + \Delta y\,\hat{y} + \Delta z\,\hat{z}$. The components of the displacement vector $\vec{s}$ are $\Delta x = 3$, $\Delta y = -8$, and $\Delta z = 0$.

Use the component form of the scalar product.

$$W = \vec{F} \cdot \vec{s} = F_x\Delta x + F_y\Delta y + F_z\Delta z = (2)(3) + (6)(-8) + (0)(0) = 6 - 48 + 0 = -42 \text{ J}$$

The work done is $W = -42$ J.

Example 123. A force $\vec{F} = 7\,\hat{x} - 9\,\hat{y} + 5\,\hat{z}$ causes the displacement $\vec{s} = 6\,\hat{x} - \hat{z}$, where SI units have been suppressed. Find the work done.

Solution. For a constant force, work is the **scalar product** (Chapter 20) between force and displacement. Identify the components of $\vec{F}$ and $\vec{s}$.

- Compare $\vec{F} = 7\,\hat{x} - 9\,\hat{y} + 5\,\hat{z}$ to $\vec{F} = F_x\hat{x} + F_y\hat{y} + F_z\hat{z}$. The components of the force vector $\vec{F}$ are $F_x = 7$, $F_y = -9$, and $F_z = 5$.
- Compare $\vec{s} = 6\,\hat{x} - \hat{z}$ to $\vec{s} = \Delta x\,\hat{x} + \Delta y\,\hat{y} + \Delta z\,\hat{z}$. The components of the displacement vector $\vec{s}$ are $\Delta x = 6$, $\Delta y = 0$, and $\Delta z = -1$. Observe that $\vec{s}$ doesn't have a $\hat{y}$: That's why $\Delta y = 0$.

Use the component form of the scalar product.

$$W = \vec{F} \cdot \vec{s} = F_x\Delta x + F_y\Delta y + F_z\Delta z = (7)(6) + (-9)(0) + (5)(-1) = 42 + 0 - 5 = 37\,\text{J}$$

The work done is $W = 37$ J.

Example 124. A monkey pulls on a box of bananas with a force of 300 N. The box of bananas travels 12 m along a straight line. The angle between the monkey's pull and the displacement is 60°. How much work does the monkey do?

Solution. Since the monkey pulls with a constant force, work is the **scalar product** (Chapter 20) between force and displacement. Use the magnitude form of the scalar product to find the work done by the monkey's pull.

$$W = F\,s\cos\theta = (300)(12)\cos 60° = 3600\left(\frac{1}{2}\right) = 1800\,\text{J} = 1.8\,\text{kJ}$$

The work done by the monkey's pull is $W = 1.8$ kJ. Recall that the metric prefix kilo (k) stands for 1000.

Example 125. As illustrated below, a monkey pulls a 20-kg box of bananas with a force of 160 N at an angle of 30° above the horizontal. The coefficient of friction between the box and the ground is $\frac{\sqrt{3}}{6}$. The box travels 7.0 m.

(A) Find the work done by the monkey's pull.

Solution. Since the monkey pulls with a constant force, work is the **scalar product** (Chapter 20) between force and displacement. Use the magnitude form of the scalar product to find the work done by the monkey's pull.

$$W = F\,s\cos\theta = (160)(7)\cos 30° = 1120\left(\frac{\sqrt{3}}{2}\right) = 560\sqrt{3}\,\text{J} = 970\,\text{J}$$

The work done by the monkey's pull is $W = 560\sqrt{3}$ J (or 970 J).

(B) Find the work done by the friction force.

Solution. Use the equation $W_{nc} = -\mu N s$ to find the work done **by friction**. (The subscripts nc represent that the work done by friction is nonconservative.) The coefficient of friction is $\mu = \frac{\sqrt{3}}{6}$. Draw a FBD to find **normal force**. (Normal force does **not** equal weight.)

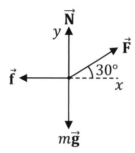

Sum the y-components of the forces to solve for normal force.

$$\sum F_y = ma_y$$

$$N + F \sin 30° - mg = 0$$

Note that $a_y = 0$ because the box doesn't accelerate vertically.

$$N = mg - F \sin 30° = (20)(9.81) - (160)\left(\frac{1}{2}\right) = 196 - 80 = 116 \text{ N}$$

Plug numbers into the equation for nonconservative work.

$$W_{nc} = -\mu N s = -\frac{\sqrt{3}}{6}(116)(7) = -234 \text{ J}$$

The work done by friction is $W_{nc} = -234 \text{ J}$ (it's $W_{nc} \approx -140\sqrt{3}$ J if you round gravity to 10 m/s^2). The work is negative because friction subtracts energy from the system.

(C) Find the work done by the normal force.

Solution. Use the equation $W = F s \cos \theta$ to find the work done by the **normal force**. When finding the work done by normal force, $\theta = 90°$ because normal force is perpendicular to the surface. Since $\cos 90° = 0$, normal force doesn't do any work: $W_N = 0$.

Example 126. A chimpanzee exerts a force $\vec{F} = 15y^4\,\hat{y}$ (where SI units have been suppressed) as a box of pineapples is displaced along a straight line from $(0, -1.0\text{ m})$ to $(0, 2.0\text{ m})$. Find the work done by the chimpanzee.

Solution. Since $\vec{F}$ is a function of y, this is a **variable force**. Therefore, we must integrate the force over the displacement in order to find the work done.

$$W = \int_{x=x_0}^{x} F_x\,dx + \int_{y=y_0}^{y} F_y\,dy$$

Compare $\vec{F} = 15y^4\,\hat{y}$ to $\vec{F} = F_x\hat{x} + F_y\hat{y} + F_z\hat{z}$. This force vector only has a y-component: $F_x = 0$, $F_y = 15y^4$, and $F_z = 0$. For this force, we just integrate over y. The limits of integration are from $y_0 = -1$ to $y = 2$ (the initial and final y-coordinates).

$$W = \int_{y=y_0}^{y} F_y\,dy = \int_{y=-1}^{2} 15y^4\,dy = \left[\frac{15y^{4+1}}{4+1}\right]_{y=-1}^{2} = \left[\frac{15y^5}{5}\right]_{y=-1}^{2} = [3y^5]_{y=-1}^{2}$$

Evaluate $3y^5$ over the limits.
$$W = 3(2)^5 - 3(-1)^5 = 3(32) - 3(-1) = 96 + 3 = 99\text{ J}$$
The work done by the chimpanzee is $W = 99$ J.

Example 127. A lemur exerts a force $\vec{F} = x\,\hat{x} + y^2\,\hat{y}$ (where SI units have been suppressed) as a box of grapes is displaced along a straight line from $(2.0\text{ m}, 3.0\text{ m})$ to $(4.0\text{ m}, 6.0\text{ m})$. Find the work done by the lemur.

Solution. Since $\vec{F}$ is a function of x and y, this is a **variable force**. Therefore, we must integrate the force over the displacement in order to find the work done.

$$W = \int_{x=x_0}^{x} F_x\,dx + \int_{y=y_0}^{y} F_y\,dy$$

Compare $\vec{F} = x\,\hat{x} + y^2\,\hat{y}$ to $\vec{F} = F_x\hat{x} + F_y\hat{y} + F_z\hat{z}$. The components of the force vector are $F_x = x$, $F_y = y^2$, and $F_z = 0$. The limits of integration are from $x_0 = 2$ to $x = 4$ and from $y_0 = 3$ to $y = 6$ (the initial and final coordinates).

$$W = \int_{x=2}^{4} x\,dx + \int_{y=3}^{6} y^2\,dy = \left[\frac{x^2}{2}\right]_{x=2}^{4} + \left[\frac{y^3}{3}\right]_{y=3}^{6}$$

$$W = \frac{(4)^2}{2} - \frac{(2)^2}{2} + \frac{(6)^3}{3} - \frac{(3)^3}{3} = \frac{16}{2} - \frac{4}{2} + \frac{216}{3} - \frac{27}{3} = 8 - 2 + 72 - 9 = 69\text{ J}$$

The work done by the lemur is $W = 69$ J.

Example 128. A gorilla exerts a force $\vec{F} = xy\,\hat{x} + x^2\,\hat{y}$ (where SI units have been suppressed) as a box of coconuts is displaced along the parabola $y = x^2$ from $(0,0)$ to $(4.0\text{ m}, 16.0\text{ m})$. Find the work done by the gorilla.

Solution. Since $\vec{F}$ is a function of x and y, this is a **variable force**. Therefore, we must integrate the force over the displacement in order to find the work done.

$$W = \int_{x=x_0}^{x} F_x\,dx + \int_{y=y_0}^{y} F_y\,dy$$

Compare $\vec{F} = xy\,\hat{x} + x^2\,\hat{y}$ to $\vec{F} = F_x\hat{x} + F_y\hat{y} + F_z\hat{z}$. The components of the force vector are $F_x = xy$, $F_y = x^2$, and $F_z = 0$. The limits of integration are from $x_0 = 0$ to $x = 4$ and from $y_0 = 0$ to $y = 16$ (the initial and final coordinates).

$$W = \int_{x=0}^{4} xy\,dx + \int_{y=0}^{16} x^2\,dy$$

We can't integrate yet because the x-integrand contains y and vice-versa. Since the box travels along the parabola $y = x^2$, we can substitute this into the x-integral. Solving for x in the equation $y = x^2$, we get $x = \sqrt{y}$, which we can substitute into the y-integral.

$$W = \int_{x=0}^{4} xy\,dx + \int_{y=0}^{16} x^2\,dy = \int_{x=0}^{4} x(x^2)\,dx + \int_{y=0}^{16} (\sqrt{y})^2\,dy = \int_{x=0}^{4} x^3\,dx + \int_{y=0}^{16} y\,dy$$

Note that $xx^2 = x^1x^2 = x^{1+2} = x^3$ according to the rule $x^m x^n = x^{m+n}$ and that $(\sqrt{y})^2 = y$.

$$W = \left[\frac{x^4}{4}\right]_{x=0}^{4} + \left[\frac{y^2}{2}\right]_{y=0}^{16} = \frac{(4)^4}{4} - \frac{(0)^4}{4} + \frac{(16)^2}{2} - \frac{(0)^2}{2} = \frac{256}{4} - 0 + \frac{256}{2} - 0 = 192\text{ J}$$

The work done by the gorilla is $W = 192$ J.

Example 129. The work done by a monkey is given by the expression $W = 72\sqrt{t}$ (where SI units have been suppressed).

(A) What is the instantaneous power at $t = 9.0$ s?

Solution. <u>Instantaneous</u> power is the <u>**derivative**</u> of work with respect to time.

$$P = \frac{dW}{dt} = \frac{d}{dt}\left(72\sqrt{t}\right)$$

Apply the rule $\sqrt{t} = t^{1/2}$.

$$P = \frac{d}{dt}\left(72t^{1/2}\right) = \left(\frac{1}{2}\right)72t^{1/2-1} = 36t^{-1/2} = \frac{36}{\sqrt{t}}$$

Note that $\left(\frac{1}{2}\right)(72) = 36$ and $\frac{1}{2} - 1 = -\frac{1}{2}$. In the last step, we applied the rule $t^{-1/2} = \frac{1}{\sqrt{t}}$.

Plug in the specified time.

$$P = \frac{36}{\sqrt{t}} = \frac{36}{\sqrt{9}} = \frac{36}{3} = 12 \text{ W}$$

The instantaneous power at the specified time is $P = 12$ W.

(B) What is the average power from $t = 0$ to $t = 9.0$ s?

Solution. <u>**Average power**</u> equals work divided by time.

$$\bar{P} = \frac{W}{t} = \frac{72\sqrt{t}}{t} = \frac{72\sqrt{9}}{9} = 8(3) = 24 \text{ W}$$

The average power over the specified interval is $\bar{P} = 24$ W. It's instructive to compare the solution for <u>**instantaneous**</u> power in part (A) to the solution for <u>**average**</u> power in part (B).

Example 130. A monkey exerts a force $\vec{\mathbf{F}} = 8\,\hat{\mathbf{x}} - 2\,\hat{\mathbf{y}}$ on a box that has an average velocity of $\vec{\bar{v}} = 6\,\hat{\mathbf{x}} - \hat{\mathbf{y}}$, where SI units have been suppressed. Find the average power delivered by the monkey.

Solution. For a constant force, average power is the <u>**scalar product**</u> (Chapter 20) between force and average velocity. Identify the components of $\vec{\mathbf{F}}$ and $\vec{\bar{v}}$.

- Compare $\vec{\mathbf{F}} = 8\,\hat{\mathbf{x}} - 2\,\hat{\mathbf{y}}$ to $\vec{\mathbf{F}} = F_x\hat{\mathbf{x}} + F_y\hat{\mathbf{y}} + F_z\hat{\mathbf{z}}$. The components of the force vector $\vec{\mathbf{F}}$ are $F_x = 8$, $F_y = -2$, and $F_z = 0$.
- Compare $\vec{\bar{v}} = 6\,\hat{\mathbf{x}} - \hat{\mathbf{y}}$ to $\vec{\bar{v}} = \bar{v}_x\,\hat{\mathbf{x}} + \bar{v}_y\,\hat{\mathbf{y}} + \bar{v}_z\,\hat{\mathbf{z}}$. The components of the average velocity $\vec{\bar{v}}$ are $\bar{v}_x = 6$, $\bar{v}_y = -1$, and $\bar{v}_z = 0$.

Use the component form of the scalar product.

$$\bar{P} = \vec{\mathbf{F}} \cdot \vec{\bar{v}} = F_x\bar{v}_x + F_y\bar{v}_y + F_z\bar{v}_z = (8)(6) + (-2)(-1) + (0)(0) = 48 + 2 + 0 = 50 \text{ W}$$

The average power delivered by the monkey is $\bar{P} = 50$ W.

Example 131. A 200-kg bananamobile uniformly accelerates from rest to 60 m/s in 5.0 s. Find the average power delivered.

Solution. Average power equals work divided by time: $\bar{P} = \frac{W}{t}$. We need to find the net work done before we can find the average power. Use the equation $W_{net} = ma_x\Delta x$ to find the **net work**. Use the equations of **uniform acceleration** (Chapter 3) to find a_x and Δx. The three known quantities for the equations of uniform acceleration are $v_{x0} = 0$, $v_x = 60$ m/s, and $t = 5.0$ s.

$$v_x = v_{x0} + a_x t$$
$$60 = 0 + a_x 5$$
$$a_x = \frac{60}{5} = 12 \text{ m/s}^2$$

Next use an equation of uniform acceleration to find the net displacement.

$$\Delta x = v_{x0}t + \frac{1}{2}a_x t^2 = 0 + \frac{1}{2}(12)(5)^2 = 6(25) = 150 \text{ m}$$

Plug numbers into the equation for the net work.

$$W_{net} = ma_x\Delta x = (200)(12)(150) = 360,000 \text{ J}$$

Now apply the equation for average power.

$$\bar{P} = \frac{W}{t} = \frac{360,000}{5} = 72,000 \text{ W}$$

The average power is $\bar{P} = 72$ kW. This is the same as 72,000 W. Recall that the metric prefix kilo (k) stands for 1000.

22 CONSERVATION OF ENERGY

Conservation of Energy

$$PE_0 + KE_0 + W_{nc} = PE + KE$$

Gravitational Potential Energy (small change in altitude)

$$PE_{g0} = mgh_0 \quad , \quad PE_g = mgh$$

Gravitational Potential Energy (astronomical change in altitude)

$$PE_{g0} = -G\frac{m_p m}{R_0} \quad , \quad PE_g = -G\frac{m_p m}{R}$$

Spring Potential Energy

$$PE_{s0} = \frac{1}{2}kx_0^2 \quad , \quad PE_s = \frac{1}{2}kx^2$$

Kinetic Energy

$$KE_0 = \frac{1}{2}mv_0^2 \quad , \quad KE = \frac{1}{2}mv^2$$

Work Done by Friction

$$W_{nc} = -\mu N s$$

Work-Energy Theorem

$$W_{net} = \Delta KE = KE - KE_0$$

Symbol	Name	SI Units
W_{net}	net work	J
PE_0	initial potential energy	J
PE	final potential energy	J
KE_0	initial kinetic energy	J
KE	final kinetic energy	J
W_{nc}	nonconservative work	J
m	mass	kg
g	gravitational acceleration	m/s^2
h_0	initial height (relative to the reference height)	m
h	final height (relative to the reference height)	m
v_0	initial speed	m/s
v	final speed	m/s
G	gravitational constant	$\frac{N \cdot m^2}{kg^2}$ or $\frac{m^3}{kg \cdot s^2}$
m_p	mass of large astronomical body	kg
R	distance from the center of the planet	m
k	spring constant	N/m or kg/s^2
x	displacement of a spring from equilibrium	m

Example 132. From the edge of an 80-m tall cliff, a monkey throws your textbook with an initial speed of 30 m/s at an angle of 30° above the horizontal. Your textbook lands on horizontal ground below. Determine the speed of your textbook just before impact.

Solution. Begin with a labeled diagram. Label the initial (i) position just after launch and the final position (f) just before impact. Put the reference height (RH) on the ground.

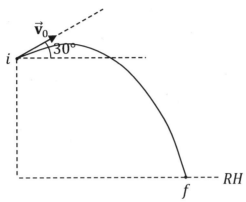

Consider the terms of the conservation of energy equation.

- There is initial gravitational potential energy ($PE_0 = mgh_0$) because the textbook is above the reference height (RH) at i.

- There is initial kinetic energy ($KE_0 = \frac{1}{2}mv_0^2$) because the textbook is moving at i.

- $W_{nc} = 0$ because there are no frictional forces (since we neglect air resistance unless stated otherwise).

- There is no final potential energy ($PE = 0$) since f is at the same height as RH (and since there isn't a spring involved in this problem).

- There is final kinetic energy ($KE = \frac{1}{2}mv^2$) because the textbook is moving at f.

  (Note that the final position is just **before** impact, not after impact.)

Write out conservation of energy for the textbook.

$$PE_0 + KE_0 + W_{nc} = PE + KE$$

Substitute the previous expressions into the conservation of energy equation.

$$mgh_0 + \frac{1}{2}mv_0^2 + 0 = 0 + \frac{1}{2}mv^2$$

Divide both sides of the equation by mass. The mass cancels out.

$$gh_0 + \frac{1}{2}v_0^2 = \frac{1}{2}v^2$$

Multiply both sides of the equation by 2.

$$2gh_0 + v_0^2 = v^2$$

Squareroot both sides of the equation.

$$v = \sqrt{2gh_0 + v_0^2} = \sqrt{2(9.81)(80) + (30)^2} = 50 \text{ m/s}$$

The final speed of the textbook just before impact is $v = 50$ m/s.

Example 133. As shown below, a box of bananas has an initial speed of 40 m/s. The box of bananas travels along the frictionless surface, which curves up a hill. What is the maximum height of the box of bananas?

Solution. Begin by labeling the diagram. Put the reference height (RH) at the bottom. Consider the terms of the conservation of energy equation.

- There is no initial potential energy ($PE_0 = 0$) since i is at the same height as RH (and since there isn't a spring involved in this problem).
- There is initial kinetic energy ($KE_0 = \frac{1}{2}mv_0^2$) because the box is moving at i.
- $W_{nc} = 0$ since the hill is frictionless.
- There is final gravitational potential energy ($PE = mgh$) because the box is above the reference height (RH) at f.
- There is no final kinetic energy ($KE = 0$) because the box runs out of speed at f (otherwise, the box would continue to rise higher). Note that the final position (f) is where the box reaches its highest point.

Write out conservation of energy for the box of bananas.

$$PE_0 + KE_0 + W_{nc} = PE + KE$$

Substitute the previous expressions into the conservation of energy equation.

$$0 + \frac{1}{2}mv_0^2 + 0 = mgh + 0$$

Divide both sides of the equation by mg. The mass cancels out.

$$h = \frac{v_0^2}{2g} = \frac{(40)^2}{2(9.81)} = 82 \text{ m}$$

The maximum height is $h = 82$ m.

Example 134. A monkey makes a pendulum by tying a 10-m long cord to a banana. The pendulum swings back and forth, beginning from rest at the leftmost position in the diagram below.

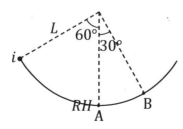

Note that the reference height (RH) is at the bottom of the arc (at point A).

(A) Determine the speed of the banana at point A.

Solution. Consider the terms of the conservation of energy equation.

- There is initial gravitational potential energy $(PE_0 = mgh_0)$ because the banana is above the reference height (RH) at i.
- There is no initial kinetic energy $(KE_0 = 0)$ because the banana begins from rest.
- $W_{nc} = 0$ neglecting any frictional forces.
- There is no final potential energy $(PE_A = 0)$ since A is at the same height as RH (and since there isn't a spring involved in this problem).
- There is final kinetic energy $(KE_A = \frac{1}{2}mv_A^2)$ because the banana is moving at A. (Point A is where the banana is moving fastest.)

Write out conservation of energy for the banana.
$$PE_0 + KE_0 + W_{nc} = PE_A + KE_A$$
Substitute the previous expressions into the conservation of energy equation.
$$mgh_0 + 0 + 0 = 0 + \frac{1}{2}mv_A^2$$
Divide both sides of the equation by mass. The mass cancels out.
$$gh_0 = \frac{1}{2}v_A^2$$
Multiply both sides of the equation by 2.
$$2gh_0 = v_A^2$$
Squareroot both sides of the equation.
$$v_A = \sqrt{2gh_0}$$
The above formula involves h_0 (the initial height), but we know L (the length of the pendulum). Draw a right triangle to relate h_0 to L.

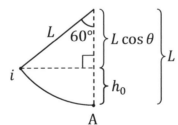

In the figure above, the two distances at the right add up to the length L (it's the radius):
$$L = h_0 + L\cos\theta$$
Subtract $L\cos\theta$ from both sides of the equation.
$$h_0 = L - L\cos\theta = 10 - 10\cos 60° = 10 - 10\left(\frac{1}{2}\right) = 10 - \frac{10}{2} = 10 - 5 = 5.0 \text{ m}$$
Plug the initial height into the equation that we previously found for speed.
$$v_A = \sqrt{2gh_0} = \sqrt{2(9.81)(5)} = \sqrt{98} = 10 \text{ m/s}$$
The speed of the banana at point A is $v_A = 10$ m/s.

(B) Determine the speed of the banana at point B.

Solution. Consider the terms of the conservation of energy equation. The final position is now at point B (instead of point A). The initial position (i) is the same.

- There is initial gravitational potential energy ($PE_0 = mgh_0$) because the banana is above the reference height (RH) at i.
- There is no initial kinetic energy ($KE_0 = 0$) because the banana begins from rest.
- $W_{nc} = 0$ neglecting any frictional forces.
- There is final potential energy ($PE_B = mgh_B$) since B is higher than RH.
- There is final kinetic energy ($KE_B = \frac{1}{2}mv_B^2$) because the banana is moving at B (because point B is below the turning point at the top of the motion).

Write out conservation of energy for the banana.

$$PE_0 + KE_0 + W_{nc} = PE_B + KE_B$$

Substitute the previous expressions into the conservation of energy equation.

$$mgh_0 + 0 + 0 = mgh_B + \frac{1}{2}mv_B^2$$

Divide both sides of the equation by mass. The mass cancels out.

$$gh_0 = gh_B + \frac{1}{2}v_B^2$$

Subtract gh_B from both sides of the equation. Factor out the g.

$$g(h_0 - h_B) = \frac{1}{2}v_B^2$$

Multiply both sides of the equation by 2.

$$2g(h_0 - h_B) = v_B^2$$

Squareroot both sides of the equation.

$$v_B = \sqrt{2g(h_0 - h_B)}$$

Solve for h_B in terms of L the same way that we did in part (A). Only the angle is different.

$$h_B = L - L\cos 30° = 10 - 10\left(\frac{\sqrt{3}}{2}\right) = 10 - 5\sqrt{3} = 1.3 \text{ m}$$

Plug the heights into the equation that we previously found for speed.

$$v_B = \sqrt{2g(h_0 - h_B)} = \sqrt{2(9.81)(5 - 1.3)} = 8.5 \text{ m/s}$$

The speed of the banana at point B is $v_B = 8.5$ m/s. (If you round 9.81 to 10, the answer can be expressed as $v_B \approx 10\sqrt{\sqrt{3} - 1}$ m/s.)

Example 135. A box of bananas begins from rest at the top of the circular hill illustrated below (the hill is exactly one quarter of a circle). The radius of the circular hill is 20 m. The hill is frictionless, but the horizontal surface is not. The coefficient of friction between the box of bananas and the <u>horizontal</u> surface is $\frac{1}{5}$. The box of bananas comes to rest at point B.

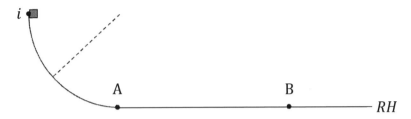

Note that the reference height (RH) is at the bottom of the hill (at both points A and B).

(A) Determine the speed of the box of bananas at point A.

Solution. Consider the terms of the conservation of energy equation.

- There is initial gravitational potential energy ($PE_0 = mgh_0$) because the box is above the reference height (RH) at i.
- There is no initial kinetic energy ($KE_0 = 0$) because the box begins from rest.
- $W_{nc} = 0$ because there isn't friction between points i and A. (Only the <u>horizontal</u> surface has friction.)
- There is no final potential energy ($PE_A = 0$) since A is at the same height as RH (and since there isn't a spring involved in this problem).
- There is final kinetic energy ($KE_A = \frac{1}{2}mv_A^2$) because the box is moving at A.

Write out conservation of energy for the box of bananas.

$$PE_0 + KE_0 + W_{nc} = PE_A + KE_A$$

Substitute the previous expressions into the conservation of energy equation.

$$mgh_0 + 0 + 0 = 0 + \frac{1}{2}mv_A^2$$

Divide both sides of the equation by mass. The mass cancels out.

$$gh_0 = \frac{1}{2}v_A^2$$

Multiply both sides of the equation by 2.

$$2gh_0 = v_A^2$$

Squareroot both sides of the equation.

$$v_A = \sqrt{2gh_0} = \sqrt{2(9.81)(20)} = \sqrt{392} = 20 \text{ m/s}$$

The speed of the box at point A is $v_A = 20$ m/s.

(B) Determine the distance between points A and B.

Solution. Consider the terms of the conservation of energy equation. The final position is now at point B (instead of point A). We may choose the initial position (i) to be the same.

- There is initial gravitational potential energy ($PE_0 = mgh_0$) because the box is above the reference height (RH) at i.
- There is no initial kinetic energy ($KE_0 = 0$) because the box begins from rest.
- $W_{nc} = -\mu Ns$ due to friction along the horizontal. Here, s is the distance between points A and B (since that's the part of the trip where there is friction).
- There is no final potential energy ($PE_B = 0$) since B is at the same height as RH (and since there isn't a spring involved in this problem).
- There is no final kinetic energy ($KE_B = 0$) because the box comes to rest at B.

Write out conservation of energy for the box of bananas.

$$PE_0 + KE_0 + W_{nc} = PE_B + KE_B$$

Substitute the previous expressions into the conservation of energy equation.

$$mgh_0 + 0 - \mu Ns = 0 + 0$$
$$mgh_0 = \mu Ns$$

Draw a FBD to find **normal force**.

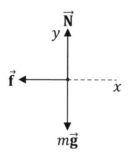

Sum the y-components of the forces to solve for the normal force.

$$\sum F_y = ma_y$$
$$N - mg = 0$$

Note that $a_y = 0$ because the box doesn't accelerate vertically. Solve for normal force in the above equation.

$$N = mg$$

Plug the expression for normal force into the equation prior to the FBD.

$$mgh_0 = \mu mgs$$

Divide both sides of the equation by mg. Mass and gravity cancel out.

$$s = \frac{h_0}{\mu} = \frac{20}{\frac{1}{5}} = 20(5) = 100 \text{ m}$$

To divide by a fraction, multiply by its **reciprocal**. Note that the reciprocal of $\frac{1}{5}$ is 5. The distance between points A and B is $s = 100$ m.

Example 136. A box of bananas slides $240\sqrt{2}$ m down a $45°$ incline from rest. The coefficient of friction between the box of bananas and the incline is $\frac{1}{4}$. Determine the speed of the box of bananas as it reaches the bottom of the incline.

Solution. Begin with a labeled diagram. Put the reference height (RH) at the bottom.

Consider the terms of the conservation of energy equation.

- There is initial gravitational potential energy $(PE_0 = mgh_0)$ because the box is above the reference height (RH) at i.
- There is no initial kinetic energy $(KE_0 = 0)$ because the box begins from rest.
- $W_{nc} = -\mu Ns$ due to friction. Here, s is the hypotenuse of the right triangle.
- There is no final potential energy $(PE = 0)$ since f is at the same height as RH (and since there isn't a spring involved in this problem).
- There is final kinetic energy $(KE = \frac{1}{2}mv^2)$ because the box is moving at f.

Write out conservation of energy for the box of bananas.

$$PE_0 + KE_0 + W_{nc} = PE + KE$$

Substitute the previous expressions into the conservation of energy equation.

$$mgh_0 + 0 - \mu Ns = 0 + \frac{1}{2}mv^2$$

Draw a FBD to find **normal force**.

Sum the y-components of the forces to solve for the normal force.

$$\sum F_y = ma_y$$

$$N - mg\cos 45° = 0$$

Note that $a_y = 0$ because the box doesn't accelerate along y (which is perpendicular to the incline); the box accelerates along x. Solve for normal force in the above equation.

$$N = mg\cos 45° = \frac{mg\sqrt{2}}{2}$$

Plug the expression for normal force into the equation prior to the FBD.

$$mgh_0 - \mu\left(\frac{mg\sqrt{2}}{2}\right)s = \frac{1}{2}mv^2$$

Divide both sides of the equation by mass. The mass cancels out.

$$gh_0 - \frac{\mu g s \sqrt{2}}{2} = \frac{1}{2}v^2$$

Multiply both sides of the equation by 2.

$$2gh_0 - \mu g s \sqrt{2} = v^2$$

Squareroot both sides of the equation.

$$v = \sqrt{2gh_0 - \mu g s \sqrt{2}}$$

The problem gives us the distance traveled: $s = 240\sqrt{2}$ m. Apply trig to solve for the initial height (h_0). Consider the right triangle below.

The side h_0 is opposite to $45°$ and s is the hypotenuse.

$$\sin 45° = \frac{h_0}{s}$$

Multiply both sides of the equation by s.

$$h_0 = s \sin 45° = \left(240\sqrt{2}\right)\left(\frac{\sqrt{2}}{2}\right) = \frac{240}{2}\sqrt{2}\sqrt{2} = 120(2) = 240 \text{ m}$$

Note that $\sqrt{2}\sqrt{2} = 2$. Plug numbers into the equation for speed that we found earlier.

$$v = \sqrt{2gh_0 - \mu g s \sqrt{2}} = \sqrt{(2)(9.81)(240) - \left(\frac{1}{4}\right)(9.81)\left(240\sqrt{2}\right)\left(\sqrt{2}\right)}$$

$$v = \sqrt{4709 - 1177} = \sqrt{3532} = 59 \text{ m/s}$$

The final speed is $v = 59$ m/s. (You get $v \approx \sqrt{3600} = 60$ m/s if you round 9.81 to 10.)

The Loop-the-Loop Roller Coaster Problem: a Prelude to Example 137. As illustrated below, a box of bananas slides from rest down a curved hill, passes once through the circular loop, and then continues along the horizontal. Neglecting friction, what is the minimum initial height needed to ensure that the box of bananas makes it safely through the loop?

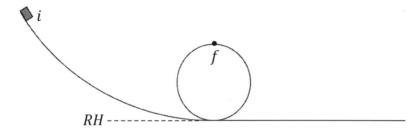

The trick to this problem is to realize that the wording "makes it safely through the loop" demands that the normal force be positive. As long as there is normal force, the box of bananas will be in contact with the surface. If the box of bananas loses contact with the surface, the normal force will be zero. Therefore, we will solve this problem by demanding that normal force be positive at the top of the loop: If the box of bananas reaches the top of the loop safely, its inertia will carry it on through. As always, the way to find normal force is to draw a FBD. Normal force pushes down (perpendicular to the surface) because the box reaches the top on the inside of the loop. Since the box is traveling in a circle, we must follow the prescription from Chapter 17 and work with inward (in) and tangential (tan) directions (not x and y).

Applying Newton's second law, where the acceleration is centripetal (a_c), we obtain:

$$\sum F_{in} = ma_c$$
$$N + mg = ma_c$$
$$N = m(a_c - g)$$

If the centripetal acceleration is greater than g, the normal force will be positive.

$$N > 0 \text{ implies that } a_c > g$$

Recall the equation for centripetal acceleration: $a_c = \frac{v^2}{R}$.

$$a_c > g$$
$$\frac{v^2}{R} > g$$
$$v^2 > Rg$$

Now we use conservation of energy to relate the initial height to the speed at the top of the loop. We'll save the above inequality for later.

Begin by drawing the path and labeling the initial position (i), final position (f), and reference height (RH). (This is drawn on the previous page.) The initial position (i) is where it starts, while the final position (f) is at the top of the loop. We choose the reference height (RH) to be at the horizontal. Consider the terms of the conservation of energy equation.

- There is initial gravitational potential energy ($PE_0 = mgh_0$) because the box is above the reference height (RH) at i.
- There is no initial kinetic energy ($KE_0 = 0$) because the box begins from rest.
- $W_{nc} = 0$ since the problem states to neglect friction.
- There is final potential energy ($PE = mgh$) since f is above RH.
- There is final kinetic energy ($KE = \frac{1}{2}mv^2$) because the box is moving at f.

Write out conservation of energy for the box of bananas.

$$PE_0 + KE_0 + W_{nc} = PE + KE$$

Substitute the previous expressions into the conservation of energy equation.

$$mgh_0 + 0 + 0 = mgh + \frac{1}{2}mv^2$$

$$gh_0 = gh + \frac{v^2}{2}$$

Note that $h = 2R$ since f is one diameter above the reference height:

$$gh_0 = g2R + \frac{v^2}{2}$$

Recall from our normal force constraint that $v^2 > Rg$. Using this, we get:

$$gh_0 > g2R + \frac{Rg}{2}$$

$$h_0 > 2R + \frac{R}{2}$$

Add fractions with a **common denominator**.

$$h_0 > \frac{4R}{2} + \frac{R}{2}$$

$$h_0 > \frac{5R}{2}$$

As long as the initial height is at least 2.5 times the radius (because $\frac{5}{2} = 2.5$), the box of bananas will make it safely through the loop, meaning that it won't lose contact with the surface at any time.

Example 137. A monkey rides the roller coaster illustrated below. The initial speed of the roller coaster is 50 m/s. The total mass of the roller coaster is 210 kg. The diameter of the loop is 35 m. Neglect friction.

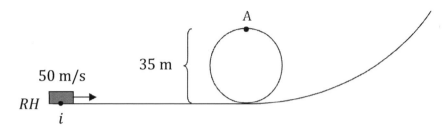

Note that the reference height (RH) is on the horizontal surface.

(A) How fast is the roller coaster moving at point A?

Solution. Consider the terms of the conservation of energy equation.

- There is no initial potential energy ($PE_0 = 0$) since i is at the same height as RH (and since there isn't a spring involved in this problem).
- There is initial kinetic energy ($KE_0 = \frac{1}{2}mv_0^2$) because the coaster is moving at i.
- $W_{nc} = 0$ since the problem states to neglect friction.
- There is final gravitational potential energy ($PE_A = mgh_A$) because the coaster is above the reference height (RH) at A.
- There is final kinetic energy ($KE_A = \frac{1}{2}mv_A^2$) because the coaster is moving at A.

Write out conservation of energy for the roller coaster.

$$PE_0 + KE_0 + W_{nc} = PE_A + KE_A$$

Substitute the previous expressions into the conservation of energy equation.

$$0 + \frac{1}{2}mv_0^2 + 0 = mgh_A + \frac{1}{2}mv_A^2$$

Divide both sides of the equation by mass. The mass cancels out.

$$\frac{1}{2}v_0^2 = gh_A + \frac{1}{2}v_A^2$$

Subtract gh_A from both sides of the equation.

$$\frac{1}{2}v_0^2 - gh_A = \frac{1}{2}v_A^2$$

Multiply both sides of the equation by 2.

$$v_0^2 - 2gh_A = v_A^2$$

Squareroot both sides of the equation.

$$v_A = \sqrt{v_0^2 - 2gh_A}$$

Note that the height at point A is equal to the diameter: $h_A = 35$ m.

$$v_A = \sqrt{(50)^2 - 2(9.81)(35)} = \sqrt{2500 - 687} = \sqrt{1813} = 43 \text{ m/s}$$

The speed at point A is $v = 43$ m/s. (You can get $v \approx \sqrt{1800} = 30\sqrt{2}$ m/s by rounding gravity from 9.81 to 10.)

161

(B) What normal force is exerted on the roller coaster at point A?

Solution. Draw a FBD at the top of the loop to find the normal force at point A.

Since the roller coaster is traveling in a circle at point A, sum the inward components of the forces to solve for normal force (like we did in Chapter 17).

$$\sum F_{in} = ma_c$$

Normal force and weight both pull toward the center of the circle, so both are positive.

$$N + mg = ma_c$$

Use the equation for centripetal acceleration: $a_c = \frac{v_A^2}{R}$. Plug this expression into the previous equation.

$$N + mg = \frac{mv_A^2}{R}$$

Subtract mg from both sides of the equation. Factor out the mass.

$$N = m\left(\frac{v_A^2}{R} - g\right) = (210)\left[\frac{(43)^2}{17.5} - 9.81\right] = (210)(106 - 9.81) = (210)(96) = 20{,}160\text{N}$$

The normal force at the top of the loop is $N = 20.2$ kN, which is the same as 20,160 N since the metric prefix kilo (k) stands for 1000. (If you round 9.81 to 10, you get $N \approx 19.5$ kN.)

Example 138. A monkey uses a spring to launch a pellet with a mass of $\frac{1}{200}$ kg straight upward. The spring is compressed $\frac{1}{8}$ m from equilibrium and released from rest. The pellet rises to a maximum height of 45 m above its initial position (the initial position is where the spring is fully compressed). What is the spring constant?

Solution. Begin with a labeled diagram. Put i at the fully compressed (FC) position and f at the top of the trajectory. Put RH at i (the fully compressed position).

Consider the terms of the conservation of energy equation.

- There is initial **spring** potential energy $(PE_{s0} = \frac{1}{2}kx_0^2)$ because the spring is compressed from equilibrium at i. There is no initial gravitational potential energy $(PE_{g0} = 0)$ since i is at the same height as RH. The total initial potential energy is $PE_0 = PE_{g0} + PE_{s0} = 0 + \frac{1}{2}kx_0^2 = \frac{1}{2}kx_0^2$.
- There is no initial kinetic energy $(KE_0 = 0)$ because the pellet begins from rest.
- $W_{nc} = 0$ because there are no frictional forces (since we neglect air resistance unless stated otherwise).
- There is final gravitational potential energy $(PE = mgh)$ because the pellet is above the reference height (RH) at f, but there is no spring potential energy since the pellet is no longer touching to the spring.
- There is no final kinetic energy $(KE = 0)$ because the pellet runs out of speed at f (otherwise, the pellet would continue to rise higher). Note that the final position (f) is where the pellet reaches its highest point.

Write out conservation of energy for the pellet.

$$PE_0 + KE_0 + W_{nc} = PE + KE$$

Substitute the previous expressions into the conservation of energy equation.

$$\frac{1}{2}kx_0^2 + 0 + 0 = mgh + 0$$

Multiply both sides of the equation by 2 and divide both sides of the equation by x_0^2.

$$k = \frac{2mgh}{x_0^2} = \frac{(2)\left(\frac{1}{200}\right)(9.81)(45)}{\left(\frac{1}{8}\right)^2} = \frac{4.4145}{\frac{1}{64}} = (4.4145)(64) = 283 \text{ N/m}$$

To divide by a fraction, multiply by its **reciprocal**. Note that the reciprocal of $\frac{1}{64}$ is 64. The spring constant is $k = 283$ N/m.

Example 139. One end of a horizontal 75 N/m spring is fixed to a vertical wall, while a 3.0-kg box of banana-shaped chocolates is connected to its free end. There is no friction between the box and the horizontal. The spring is compressed 5.0-m from the equilibrium position and released from rest.

Begin by labeling the diagram. Put the initial position (i) at the fully compressed (FC) position, the final position (f_A) for part (A) at the equilibrium (EQ) position, and the final position (f_B) for part (B) between EQ and the fully stretched (FS) position. Put the reference height (RH) on the horizontal surface.

(A) What is the speed of the box when the system passes through equilibrium?

Solution. Consider the terms of the conservation of energy equation.

- There is initial **spring** potential energy $(PE_{s0} = \frac{1}{2}kx_0^2)$ because the spring is compressed from equilibrium at i. There is no initial gravitational potential energy $(PE_{g0} = 0)$ since i is at the same height as RH. The total initial potential energy is $PE_0 = PE_{g0} + PE_{s0} = 0 + \frac{1}{2}kx_0^2 = \frac{1}{2}kx_0^2$.

- There is no initial kinetic energy $(KE_0 = 0)$ because the box begins from rest.

- $W_{nc} = 0$ because there is no friction.

- There is no final **spring** potential energy $(PE_{sA} = 0)$ because the spring is at equilibrium at f_A. There is no final gravitational potential energy $(PE_{gA} = 0)$ since f_A is at the same height as RH. The total final potential energy is $PE_A = PE_{gA} + PE_{sA} = 0 + 0 = 0$.

- There is final kinetic energy $(KE_A = \frac{1}{2}mv_A^2)$ because the box is moving at f_A. (As the box oscillates back and forth, it moves fastest when passing through equilibrium.)

Write out conservation of energy for the box of chocolates.

$$PE_0 + KE_0 + W_{nc} = PE_A + KE_A$$

Substitute the previous expressions into the conservation of energy equation.

$$\frac{1}{2}kx_0^2 + 0 + 0 = 0 + \frac{1}{2}mv_A^2$$

Multiply both sides of the equation by 2.

$$kx_0^2 = mv_A^2$$

Divide both sides of the equation by mass.

$$v_A^2 = \frac{kx_0^2}{m}$$

Squareroot both sides of the equation. Note that $\sqrt{x_0^2} = x_0$.

$$v_A = x_0\sqrt{\frac{k}{m}} = 5\sqrt{\frac{75}{3}} = 5\sqrt{25} = 5(5) = 25 \text{ m/s}$$

The speed of the box as it passes through the equilibrium position is $v_A = 25$ m/s.

(B) What is the speed of the box when the spring is stretched 4.0 m from equilibrium?

Solution. Consider the terms of the conservation of energy equation. Note that the final position is now between equilibrium (EQ) and the fully stretched (FS) position.

- There is initial **spring** potential energy $(PE_{s0} = \frac{1}{2}kx_0^2)$ because the spring is compressed from equilibrium at i. There is no initial gravitational potential energy $(PE_{g0} = 0)$ since i is at the same height as RH. The total initial potential energy is $PE_0 = PE_{g0} + PE_{s0} = 0 + \frac{1}{2}kx_0^2 = \frac{1}{2}kx_0^2$.

- There is no initial kinetic energy $(KE_0 = 0)$ because the box begins from rest.

- $W_{nc} = 0$ because there is no friction.

- There is final **spring** potential energy ($PE_{SB} = \frac{1}{2}kx_B^2$) because the spring is stretched from equilibrium at f_B. There is no final gravitational potential energy ($PE_{gB} = 0$) since f_B is at the same height as RH. The total final potential energy is $PE_B = PE_{gB} + PE_{SB} = 0 + \frac{1}{2}kx_B^2 = \frac{1}{2}kx_B^2$.

- There is final kinetic energy ($KE_B = \frac{1}{2}mv_B^2$) because the box is moving at f_B. (The box only comes to rest at the two endpoints of the motion, at FC and FS.)

Write out conservation of energy for the box of chocolates.

$$PE_0 + KE_0 + W_{nc} = PE_B + KE_B$$

Substitute the previous expressions into the conservation of energy equation.

$$\frac{1}{2}kx_0^2 + 0 + 0 = \frac{1}{2}kx_B^2 + \frac{1}{2}mv_B^2$$

Multiply both sides of the equation by 2.

$$kx_0^2 = kx_B^2 + mv_B^2$$

Subtract kx_B^2 from both sides of the equation. Factor out the spring constant.

$$k(x_0^2 - x_B^2) = mv_B^2$$

Divide both sides of the equation by mass.

$$v_B^2 = \frac{k}{m}(x_0^2 - x_B^2)$$

Squareroot both sides of the equation.

$$v_B = \sqrt{\frac{k}{m}(x_0^2 - x_B^2)} = \sqrt{\frac{75}{3}(5^2 - 4^2)} = \sqrt{(25)(25 - 16)} = \sqrt{(25)(9)} = (5)(3) = 15 \text{ m/s}$$

The speed of the box at point B is $v_B = 15$ m/s.

Example 140. Planet FurryTail has a mass of 1.5×10^{26} kg and a radius of 2.0×10^6 m. What is the escape speed for a projectile leaving the surface of FurryTail?

Solution. Begin by drawing the path and labeling the initial position (i), final position (f), and reference height (RH). The initial position (i) is on the surface of FurryTail, while the final position (f) and reference height (RH) are infinitely far away.

Why infinitely far away? That's how far away the projectile would eventually need to get in order to completely escape the influence of FurryTail's gravitational pull. The reference height (RH) is the place where gravitational potential energy equals zero. For a significant change in altitude, we work with the expression $PE_g = -\frac{Gm_p m}{R}$, so the reference height is the place where R is infinite. (Note that $\frac{1}{R}$ approaches zero as R approaches infinity.)

Consider the terms of the conservation of energy equation.

- There is initial gravitational potential energy ($PE_0 = -\frac{Gm_pm}{R_0}$) because i is not at the reference height (RH). We must use the equation $PE_0 = -\frac{Gm_pm}{R_0}$ (instead of mgh_0) for an **astronomical** change in altitude.
- There is initial kinetic energy ($KE_0 = \frac{1}{2}mv_0^2$) because the projectile is moving at i.
- $W_{nc} = 0$ since we neglect air resistance unless stated otherwise. (Realistically, air resistance would be significant as the projectile travels through the atmosphere, before it enters space.)
- There is no final potential energy ($PE = 0$) since f is at the same position as RH (and since there isn't a spring involved in this problem). Note that the final position is infinitely far away from the planet (that's how far the projectile needs to travel in order to completely escape the influence of the planet's gravitational field).
- There is no final kinetic energy ($KE = 0$) if the projectile has the minimum speed needed to reach the final position (if the projectile has any extra initial speed, then it would be moving when it reaches the final position; but we're trying to calculate the **minimum** speed needed to escape the planet's gravitational pull).

Write out conservation of energy for the projectile.

$$PE_0 + KE_0 + W_{nc} = PE + KE$$

Substitute the previous expressions into the conservation of energy equation.

$$-\frac{Gm_pm}{R_0} + \frac{1}{2}mv_0^2 + 0 = 0 + 0$$

Add $\frac{Gm_pm}{R_0}$ to both sides of the equation.

$$\frac{1}{2}mv_0^2 = \frac{Gm_pm}{R_0}$$

Divide both sides of the equation by the mass of the projectile (m). The projectile's mass cancels out. However, the mass of the planet (m_p) does not cancel out.

$$\frac{v_0^2}{2} = \frac{Gm_p}{R_0}$$

Multiply both sides of the equation by 2.

$$v_0^2 = \frac{2Gm_p}{R_0}$$

Squareroot both sides of the equation.

$$v_0 = \sqrt{\frac{2Gm_p}{R_0}} = \sqrt{\frac{2\left(\frac{2}{3} \times 10^{-10}\right)(1.5 \times 10^{26})}{2 \times 10^6}} = \sqrt{\frac{1 \times 10^{16}}{10^6}} = \sqrt{1 \times 10^{10}} = 1.0 \times 10^5 \text{ m/s}$$

The escape speed is $v = 100$ km/s, which is the same as $v = 10^5$ m/s since the prefix kilo (k) stands for 1000.

Example 141. Planet SillyMonk has a mass of 3.0×10^{24} kg. A 2000-kg satellite traveling in an elliptical orbit is moving $2000\sqrt{19}$ m/s when it is 8.0×10^7 m from the center of Silly-Monk. What is the speed of the satellite when it is 4.0×10^7 m from the center of Silly-Monk?

Solution. Begin by drawing the path and labeling the initial position (i), final position (f), and reference height (RH). The initial position (i) is where $R_0 = 8.0 \times 10^7$ m, the final position (f) is where $R = 4.0 \times 10^7$ m, and the reference height (RH) is infinitely far away (as we discussed in the prior example).

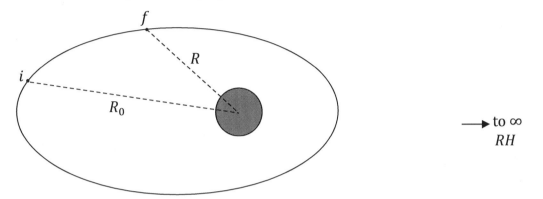

Consider the terms of the conservation of energy equation.

- There is initial gravitational potential energy $(PE_0 = -\frac{Gm_p m}{R_0})$ because i is not at the reference height (RH). We must use the equation $PE_0 = -\frac{Gm_p m}{R_0}$ (instead of mgh_0) for an **astronomical** change in altitude.

- There is initial kinetic energy $(KE_0 = \frac{1}{2}mv_0^2)$ because the satellite is moving at i.

- $W_{nc} = 0$ since we neglect air resistance unless stated otherwise. (This is realistic for a satellite traveling through space.)

- There is final gravitational potential energy $(PE = -\frac{Gm_p m}{R})$ because i is not at the reference height (RH). We must use the equation $PE = -\frac{Gm_p m}{R}$ (instead of mgh) for an **astronomical** change in altitude.

- There is initial kinetic energy $(KE = \frac{1}{2}mv^2)$ because the satellite is moving at f.

Write out conservation of energy for the satellite.

$$PE_0 + KE_0 + W_{nc} = PE + KE$$

Substitute the previous expressions into the conservation of energy equation.

$$-\frac{Gm_p m}{R_0} + \frac{1}{2}mv_0^2 + 0 = -\frac{Gm_p m}{R} + \frac{1}{2}mv^2$$

Divide both sides of the equation by the mass of the satellite (m). The satellite's mass cancels out. However, the mass of the planet (m_p) does not cancel out.

$$-\frac{Gm_p}{R_0} + \frac{v_0^2}{2} = -\frac{Gm_p}{R} + \frac{v^2}{2}$$

Add $\frac{Gm_p}{R}$ to both sides of the equation.

$$\frac{Gm_p}{R} - \frac{Gm_p}{R_0} + \frac{v_0^2}{2} = \frac{v^2}{2}$$

Factor out Gm_p from the two terms on the left-hand side of the equation.

$$Gm_p \left(\frac{1}{R} - \frac{1}{R_0}\right) + \frac{v_0^2}{2} = \frac{v^2}{2}$$

Multiply both sides of the equation by 2.

$$2Gm_p \left(\frac{1}{R} - \frac{1}{R_0}\right) + v_0^2 = v^2$$

Squareroot both sides of the equation.

$$v = \sqrt{2Gm_p \left(\frac{1}{R} - \frac{1}{R_0}\right) + v_0^2}$$

$$v = \sqrt{(2)\left(\frac{2}{3} \times 10^{-10}\right)(3 \times 10^{24})\left(\frac{1}{4 \times 10^7} - \frac{1}{8 \times 10^7}\right) + \left(2000\sqrt{19}\right)^2}$$

Subtract the fractions by making a common denominator. Multiply $\frac{1}{4 \times 10^7}$ by $\frac{2}{2}$ to get $\frac{2}{8 \times 10^7}$.

$$v = \sqrt{(4 \times 10^{14})\left(\frac{2}{8 \times 10^7} - \frac{1}{8 \times 10^7}\right) + 7.6 \times 10^7}$$

$$v = \sqrt{(4 \times 10^{14})\left(\frac{1}{8 \times 10^7}\right) + 7.6 \times 10^7} = \sqrt{\frac{4 \times 10^{14}}{8 \times 10^7} + 7.6 \times 10^7}$$

$$v = \sqrt{0.5 \times 10^7 + 7.6 \times 10^7} = \sqrt{8.1 \times 10^7} = 9000 \text{ m/s}$$

The final speed is $v = 9.0$ km/s, which is the same as $v = 9000$ m/s since the prefix kilo (k) stands for 1000.

23 ONE-DIMENSIONAL COLLISIONS

Momentum
$$\vec{p} = m\vec{v}$$

Impulse
$$\vec{J} = \Delta\vec{p} = \vec{p} - \vec{p}_0 = \vec{F}_c\Delta t$$

Conservation of Momentum
$$m_1\vec{v}_{10} + m_2\vec{v}_{20} = m_1\vec{v}_1 + m_2\vec{v}_2$$

Perfectly Inelastic Collision
$$m_1\vec{v}_{10} + m_2\vec{v}_{20} = (m_1 + m_2)\vec{v}$$

Inverse Perfectly Inelastic Collision
$$(m_1 + m_2)\vec{v}_0 = m_1\vec{v}_1 + m_2\vec{v}_2$$

Elastic Collision (One-dimensional)
$$m_1\vec{v}_{10} + m_2\vec{v}_{20} = m_1\vec{v}_1 + m_2\vec{v}_2$$ $$\vec{v}_{10} + \vec{v}_1 = \vec{v}_{20} + \vec{v}_2$$

Percent of Kinetic Energy Lost or Gained
$$KE_0 = \frac{1}{2}m_1v_{10}^2 + \frac{1}{2}m_2v_{20}^2 \quad , \quad KE = \frac{1}{2}m_1v_1^2 + \frac{1}{2}m_2v_2^2$$ $$\% \text{ change} = \frac{|KE - KE_0|}{KE_0} \times 100\%$$

Symbol	Name	SI Units
$\vec{p}_0$	initial momentum	kg·m/s
$\vec{p}$	momentum	kg·m/s
m	mass	kg
$\vec{v}$	velocity	m/s
v	speed	m/s
$\vec{J}$	impulse	Ns
$\Delta\vec{p}$	change in momentum	kg·m/s
$\vec{F}_c$	average collision force	N
Δt	duration of the collision (time interval)	s
m_1	mass of object 1	kg
m_2	mass of object 2	kg
$\vec{v}_{10}$	initial velocity of object 1	m/s
$\vec{v}_{20}$	initial velocity of object 2	m/s
$\vec{v}_1$	final velocity of object 1	m/s
$\vec{v}_2$	final velocity of object 2	m/s
KE_0	initial kinetic energy of the system	J
KE	final kinetic energy of the system	J

Example 142. A monkey places a 500-g banana on top of his head. Another monkey shoots the banana with a 250-g arrow. The arrow sticks in the banana. The arrow is traveling 45 m/s horizontally just before impact. How fast do the banana and arrow travel just after impact?

Solution. First setup a coordinate system. We choose $+x$ to point along the direction that the arrow is shot. With this choice, the arrow's initial velocity ($\vec{v}_{20}$) will be along the unit vector $\hat{x}$. Recall that the unit vector $\hat{x}$ points one unit along the $+x$-axis. (We encountered unit vectors at the end of Chapter 10.) Identify the known quantities for object 1 (the banana) and object 2 (the arrow).

- The mass of the banana is $m_1 = 500$ g.
- The initial velocity of the banana is zero: $\vec{v}_{10} = 0$. The banana is initially at rest.
- The mass of the arrow is $m_2 = 250$ g.
- The initial velocity of the arrow is $\vec{v}_{20} = 45\hat{x}$ m/s.

Since the arrow and banana stick together, use the equation for a **perfectly inelastic collision**. The arrow and banana have the same final velocity.

$$m_1\vec{v}_{10} + m_2\vec{v}_{20} = (m_1 + m_2)\vec{v}$$

Note that you **don't** need to convert grams (g) to kilograms, so long as you're consistent: If every mass is in grams, the grams will cancel out.

$$(500)(0) + (250)(45\hat{x}) = (500 + 250)\vec{v}$$
$$0 + 11{,}250\hat{x} = 750\vec{v}$$
$$\vec{v} = \frac{11{,}250}{750}\hat{x} = 15\hat{x}\text{ m/s}$$

The final velocity of the banana and arrow is $\vec{v} = 15\hat{x}$ m/s, while the final speed is $v = 15$ m/s. The distinction is that the velocity includes direction: The unit vector $\hat{x}$ indicates that the banana and arrow travel along $+x$ after the collision.

Example 143. A 50-kg monkey is initially sitting in a 250-kg canoe. The canoe is initially at rest relative to the lake. The monkey begins to walk 3.0 m/s (relative to the water) to the south from one end of the canoe to the other. What is the velocity of the canoe while the monkey walks to the south?

Solution. First setup a coordinate system. We choose $+y$ to point north. With this choice, the monkey's final velocity ($\vec{v}_1$) will be along $-\hat{y}$ (since he walks to the **south**). Identify the known quantities for object 1 (the monkey) and object 2 (the canoe).

- The mass of the monkey is $m_1 = 50$ kg.
- The initial velocity of the monkey and canoe is zero: $\vec{v}_0 = 0$. They start from rest.
- The final velocity of the monkey is $\vec{v}_1 = -3\hat{y}$ m/s since it is relative to the water. (If this were relative to the canoe, we would need to perform vector subtraction with the two velocities.) The monkey's final velocity is negative because he walks **south** (while we setup our coordinate system with $+y$ pointing north).
- The mass of the canoe is $m_2 = 250$ kg.

Since the monkey and canoe have the same initial velocity, use the equation for an **inverse perfectly inelastic collision**.

$$(m_1 + m_2)\vec{v}_0 = m_1\vec{v}_1 + m_2\vec{v}_2$$
$$(50 + 250)(0) = (50)(-3\hat{y}) + 250\vec{v}_2$$
$$0 = -150\hat{y} + 250\vec{v}_2$$
$$150\hat{y} = 250\vec{v}_2$$
$$\vec{v}_2 = \frac{150}{250}\hat{y} = \frac{3}{5}\hat{y} \, \text{m/s}$$

The final velocity of the canoe is $\vec{v}_2 = \frac{3}{5}\hat{y}$ m/s, where the unit vector $\hat{y}$ indicates that the canoe travels **north** (along $+y$) after the collision. Note that $\frac{3}{5}$ can also be expressed as 0.60.

Example 144. A 6.0-kg box of bananas traveling 5.0 m/s to the east collides head-on with a 12.0-kg box of apples traveling 4.0 m/s to the west on horizontal frictionless ice.
(A) Determine the final velocity of each box if the collision is perfectly inelastic.
Solution. First setup a coordinate system. We choose $+x$ to point east. With this choice, the initial velocity ($\vec{v}_{10}$) of the box of bananas will be along $\hat{x}$, while the initial velocity ($\vec{v}_{20}$) of the box of apples will be along $-\hat{x}$. Identify the known quantities for object 1 (the box of bananas) and object 2 (the box of apples).

- The mass of the box of bananas is $m_1 = 6.0$ kg.
- The initial velocity of the box of bananas is $\vec{v}_{10} = 5\hat{x}$ m/s.
- The mass of the box of apples is $m_2 = 12.0$ kg.
- The initial velocity of the box of apples is $\vec{v}_{20} = -4\hat{x}$ m/s.

In part (A), use the equation for a **perfectly inelastic collision**. The boxes have the same final velocity.

$$m_1\vec{v}_{10} + m_2\vec{v}_{20} = (m_1 + m_2)\vec{v}$$
$$(6)(5\hat{x}) + (12)(-4\hat{x}) = (6 + 12)\vec{v}$$
$$30\hat{x} - 48\hat{x} = 18\vec{v}$$
$$-18\hat{x} = 18\vec{v}$$
$$\vec{v} = \frac{-18}{18}\hat{x} = -1.0\,\hat{x} \, \text{m/s}$$

The final velocity of the boxes is $\vec{v} = -1.0\,\hat{x}$ m/s, while the final speed is $v = 1.0$ m/s. The distinction is that the velocity includes direction: The unit vector $\hat{x}$ indicates that the boxes travel **west** (along $-x$) after the collision (since we chose $\hat{x}$ to be east).
(B) Determine the final velocity of each box if the collision is elastic.
Solution. We still have the same knowns for the initial quantities as we did in part (A), but now we must use the two equations for an **elastic collision** (instead of the equation for a perfectly inelastic collision). We will use the same coordinate system, with $+x$ pointing to the east.

$$m_1\vec{v}_{10} + m_2\vec{v}_{20} = m_1\vec{v}_1 + m_2\vec{v}_2$$
$$\vec{v}_{10} + \vec{v}_1 = \vec{v}_{20} + \vec{v}_2$$

Plug numbers into the two equations for an elastic collision.

$$(6)(5\hat{x}) + (12)(-4\hat{x}) = 6\vec{v}_1 + 12\vec{v}_2$$
$$5\hat{x} + \vec{v}_1 = -4\hat{x} + \vec{v}_2$$

Simplify the left side of the first equation and add $4\hat{x}$ to both sides of the second equation.

$$30\hat{x} - 48\hat{x} = 6\vec{v}_1 + 12\vec{v}_2$$
$$5\hat{x} + 4\hat{x} + \vec{v}_1 = \vec{v}_2$$

Simplify the left side of each equation.

$$-18\hat{x} = 6\vec{v}_1 + 12\vec{v}_2$$
$$9\hat{x} + \vec{v}_1 = \vec{v}_2$$

Since $9\hat{x} + \vec{v}_1 = \vec{v}_2$, plug $9\hat{x} + \vec{v}_1$ in for $\vec{v}_2$ in the first equation.

$$-18\hat{x} = 6\vec{v}_1 + 12(9\hat{x} + \vec{v}_1)$$

Distribute the 12 to both terms.

$$-18\hat{x} = 6\vec{v}_1 + 108\hat{x} + 12\vec{v}_1$$

Combine like terms. Subtract $108\hat{x}$ from both sides. Add $6\vec{v}_1$ and $12\vec{v}_1$ together.

$$-126\hat{x} = 18\vec{v}_1$$

Divide both sides of the equation by 18.

$$\vec{v}_1 = \frac{-126}{18}\hat{x} = -7.0\,\hat{x}\,\text{m/s}$$

Plug this answer for $\vec{v}_1$ into the equation that we previously found for $\vec{v}_2$.

$$\vec{v}_2 = 9\hat{x} + \vec{v}_1 = 9\hat{x} - 7\hat{x} = 2.0\,\hat{x}\,\text{m/s}$$

The final velocities of the boxes are $\vec{v}_1 = -7.0\,\hat{x}$ m/s and $\vec{v}_2 = 2.0\,\hat{x}$ m/s. Since we chose $+x$ to point east, the box of bananas travels **west** (since $\vec{v}_1$ is negative) and the box of apples travels **east** (since $\vec{v}_2$ is positive).

Example 145. A 3.0-kg box of coconuts traveling 16.0 m/s to the north collides head-on with a 9.0-kg box of grapefruit traveling 6.0 m/s to the south on horizontal frictionless ice. The collision is elastic. The collision lasts for a duration of 250 ms.

(A) Determine the final velocity of each box.

Solution. First setup a coordinate system. We choose $+y$ to point north. With this choice, the initial velocity ($\vec{v}_{10}$) of the box of coconuts will be along $\hat{y}$, while the initial velocity ($\vec{v}_{20}$) of the box of grapefruit will be along $-\hat{y}$. Identify the known quantities for object 1 (the box of coconuts) and object 2 (the box of grapefruit).

- The mass of the box of coconuts is $m_1 = 3.0$ kg.
- The initial velocity of the box of coconuts is $\vec{v}_{10} = 16\hat{y}$ m/s.
- The mass of the box of grapefruit is $m_2 = 9.0$ kg.
- The initial velocity of the box of grapefruit is $\vec{v}_{20} = -6\hat{y}$ m/s.
- The duration of the collision is 250 ms. Convert this to seconds, given that 1 s $= 1000$ ms (where ms stands for milliseconds). The time interval is $\Delta t = 0.25$ s.

Use the two equations for an **elastic collision**.

$$m_1\vec{v}_{10} + m_2\vec{v}_{20} = m_1\vec{v}_1 + m_2\vec{v}_2$$
$$\vec{v}_{10} + \vec{v}_1 = \vec{v}_{20} + \vec{v}_2$$

Plug numbers into the two equations for an elastic collision.

$$(3)(16\hat{y}) + (9)(-6\hat{y}) = 3\vec{v}_1 + 9\vec{v}_2$$
$$16\hat{y} + \vec{v}_1 = -6\hat{y} + \vec{v}_2$$

Simplify the left side of the first equation and add $6\hat{y}$ to both sides of the second equation.

$$48\hat{y} - 54\hat{y} = 3\vec{v}_1 + 9\vec{v}_2$$
$$16\hat{y} + 6\hat{y} + \vec{v}_1 = \vec{v}_2$$

Simplify the left side of each equation.

$$-6\hat{y} = 3\vec{v}_1 + 9\vec{v}_2$$
$$22\hat{y} + \vec{v}_1 = \vec{v}_2$$

Since $22\hat{y} + \vec{v}_1 = \vec{v}_2$, plug $22\hat{y} + \vec{v}_1$ in for $\vec{v}_2$ in the first equation.

$$-6\hat{y} = 3\vec{v}_1 + 9(22\hat{y} + \vec{v}_1)$$

Distribute the 9 to both terms.

$$-6\hat{y} = 3\vec{v}_1 + 198\hat{y} + 9\vec{v}_1$$

Combine like terms. Subtract $198\hat{y}$ from both sides. Add $3\vec{v}_1$ and $9\vec{v}_1$ together.

$$-204\hat{y} = 12\vec{v}_1$$

Divide both sides of the equation by 12.

$$\vec{v}_1 = \frac{-204}{12}\hat{y} = -17.0\,\hat{y}\,\text{m/s}$$

Plug this answer for $\vec{v}_1$ into the equation that we previously found for $\vec{v}_2$.

$$\vec{v}_2 = 22\hat{y} + \vec{v}_1 = 22\hat{y} - 17\hat{y} = 5.0\,\hat{y}\,\text{m/s}$$

The final velocities of the boxes are $\vec{v}_1 = -17.0\,\hat{y}$ m/s and $\vec{v}_2 = 5.0\,\hat{y}$ m/s. Since we chose $+y$ to point north, the box of coconuts travels **south** (since $\vec{v}_1$ is negative) and the box of grapefruit travels **north** (since $\vec{v}_2$ is positive).

(B) What is the magnitude of the average collision force?

Solution. Apply the definition of **impulse**. The impulse $(\vec{J})$ is defined as the change in momentum $(\Delta\vec{p} = \vec{p} - \vec{p}_0)$, and impulse also equals the average collision force $(\vec{F}_c)$ times the duration of the collision (Δt). Apply the impulse equations to a single object (just one box). We will choose the box of coconuts.

$$\vec{J}_1 = \Delta\vec{p}_1 = \vec{p}_1 - \vec{p}_{10} = m_1(\vec{v}_1 - \vec{v}_{10}) = \vec{F}_c\Delta t$$

Solve for the average collision force exerted on object 1.

$$\vec{F}_c = \frac{m_1(\vec{v}_1 - \vec{v}_{10})}{\Delta t} = \frac{(3)(-17\hat{y} - 16\hat{y})}{0.25} = 12(-33\hat{y}) = -396\,\hat{y}\,\text{N}$$

The magnitude of the average collision force is $F_c = 396$ N, where the $-\hat{y}$ refers to the direction (it points south for the average force exerted on the box of coconuts). The average collision force exerted on object 2 (the box of grapefruit) is equal and opposite according to Newton's third law: It equals $+396\,\hat{y}$ N.

The Ballistic Pendulum: A Prelude to Example 146. As illustrated below, a monkey shoots a bullet of mass m_b ("b" for bullet) into a block of wood of mass m_p ("p" for pendulum). The bullet is traveling horizontally just before impact. The block of wood is initially at rest. The bullet sticks inside the block of wood and the system rises upward to a maximum angle with the vertical of θ. The length of the pendulum from the pivot to the center of mass of the system is L. This is called a **ballistic pendulum**. The goal is to determine the initial speed of the bullet, just before impact.

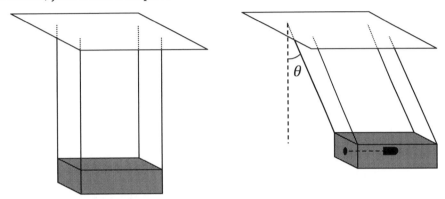

The ballistic pendulum involves more than just a collision:
- It begins with a perfectly inelastic collision at the bottom.
- After the collision, the pendulum and bullet swing together upward in an arc.

We must treat the two processes separately:
- Only momentum is conserved for the perfectly inelastic collision. (As the collision is inelastic, mechanical energy isn't conserved for this process.)
- Only mechanical energy is conserved for the swing upward. (The momentum of the system is clearly lost during the swing upward, since the system has momentum at the bottom and comes to rest at the top.)

This example gives us information about the final position (θ), and asks for the initial speed of the bullet prior to the collision. We must therefore solve this problem in reverse, beginning with the swing upward and treating the collision last.

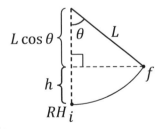

Apply conservation of energy (Chapter 22) to the swing upward. Begin by drawing the path and labeling the initial position (i), final position (f), and reference height (RH). The initial position (i) is just **after** the collision (because mechanical energy is not conserved for an inelastic collision), while the final position (f) is at the top of the arc. We choose the reference height (RH) to be at the bottom of the arc. Write out conservation of energy.

$$PE_{ac} + KE_{ac} + W_{nc} = PE + KE$$

The subscript ac stands for "after collision." The initial position of the swing upward will be the final position for the collision. It would be confusing to call the initial speed v_0 for the swing upward and then proceed to use the same symbol v_0 for the initial speed for the collision (since the initial speed for the collision is different from the initial speed for the swing upward). Consider the terms of the conservation of energy equation.

- There is no initial potential energy ($PE_{ac} = 0$) since i is at the same height as RH (and since there isn't a spring involved in this problem).
- There is initial kinetic energy, equal to $KE_{ac} = \frac{1}{2}(m_b + m_p)v_{ac}^2$, because the pendulum is moving at i.
- $W_{nc} = 0$ neglecting air resistance and friction at the point of support.
- There is final gravitational potential energy, equal to $PE = (m_b + m_p)gh$, because the pendulum is above the reference height (RH) at f.
- There is no final kinetic energy ($KE = 0$) because the pendulum runs out of speed at f (otherwise, the pendulum would continue to rise higher). Note that the final position (f) is where the pendulum reaches its highest point.

Substitute the previous expressions into the conservation of energy equation.

$$0 + \frac{1}{2}(m_b + m_p)v_{ac}^2 + 0 = (m_b + m_p)gh + 0$$

Divide both sides of the equation by $(m_b + m_p)$. Mass cancels. Multiply both sides by 2.

$$v_{ac}^2 = 2gh$$

Squareroot both sides of the equation.

$$v_{ac} = \sqrt{2gh}$$

We need to relate the length of the pendulum (L) to the final height (h). We can do this using geometry. Study the diagram below.

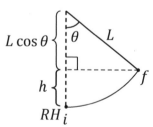

In the figure above, the two distances at the left (h and $L\cos\theta$) add up to the length L.

$$L = h + L\cos\theta$$

$$h = L - L\cos\theta = L(1 - \cos\theta)$$

Substitute this expression into the equation $v_{ac} = \sqrt{2gh}$.

$$v_{ac} = \sqrt{2gL(1 - \cos\theta)}$$

Now we are prepared to treat the collision. We choose $+x$ to be the forward direction of the bullet. Only momentum is conserved for the perfectly inelastic collision at the bottom.

$$m_b\vec{v}_{b0} + m_p\vec{v}_{p0} = (m_b + m_p)\vec{v}_{ac}$$

Note that the subscripts are b for bullet and p for pendulum. The final velocity of the collision, $\vec{v}_{ac}$, is the same as the initial velocity for the swing upward. We already found an expression for v_{ac}. The pendulum is initially at rest: $\vec{v}_{p0} = 0$.

$$m_b\vec{v}_{b0} + 0 = (m_b + m_p)\vec{v}_{ac}$$

Divide both sides by the mass of the bullet.

$$v_{b0} = \frac{m_b + m_p}{m_b} v_{ac}$$

Substitute the expression for v_{ac} that we found previously into the above equation.

$$v_{b0} = \frac{m_b + m_p}{m_b} \sqrt{2gL(1 - \cos\theta)}$$

If we measure the mass of the bullet (m_b), the mass of the pendulum (m_p), the length of the pendulum (L), and the maximum angle (θ), we can use this equation to calculate the speed (v_{b0}) of the bullet just before the collision.

Example 146. A 20-kg box of bananas begins from rest at the top of the circular hill illustrated below (the hill is exactly one quarter of a circle). The radius of the circular hill is 5.0 m. The hill and horizontal are both frictionless. When the box of bananas reaches point A, it collides with a 30-kg box of pineapples which is initially at rest. The two boxes stick together after the collision. Determine the final speed of the boxes.

Solution. Solve this problem in two stages:
- Stage 1: Apply the law of conservation of energy to solve for the speed of the box of bananas at point A just prior to the collision.
- Stage 2: Apply the law of conservation of momentum to the collision.

Stage 1: Conserve energy to find the speed of the box of bananas just prior to the collision. Put i at the top of the hill and f_{bc} ("bc" stands for "**before collision**") just before point A. Put RH on the horizontal. Consider the terms of the conservation of energy equation.
- There is initial gravitational potential energy ($PE_0 = mgh_0$) because the box is above the reference height (RH) at i.
- There is no initial kinetic energy ($KE_0 = 0$) because the box begins from rest.
- $W_{nc} = 0$ since the problem states that the surface is frictionless.
- There is no final potential energy ($PE_{bc} = 0$) since f_{bc} is at the same height as RH (and since there isn't a spring involved in this problem).

- There is final kinetic energy ($KE_{bc} = \frac{1}{2}mv_{bc}^2$) because the box is moving at f_{bc}.

Write out conservation of energy for the box of bananas.

$$PE_0 + KE_0 + W_{nc} = PE_{bc} + KE_{bc}$$

Substitute the previous expressions into the conservation of energy equation.

$$mgh_0 + 0 + 0 = 0 + \frac{1}{2}mv_{bc}^2$$

Divide both sides of the equation by m. Mass cancels. Multiply both sides by 2.

$$2gh_0 = v_{bc}^2$$

Squareroot both sides of the equation.

$$v_{bc} = \sqrt{2gh_0}$$

The initial height equals the radius of the circle: $h_0 = R = 5.0$ m.

$$v_{bc} = \sqrt{(2)(9.81)(5)} = \sqrt{98.1} = 9.9 \text{ m/s}$$

Stage 2: Conserve momentum for the **perfectly inelastic** collision. The boxes have the same final velocity because they stick together.

$$m_1\vec{v}_{bc} + m_2\vec{v}_{20} = (m_1 + m_2)\vec{v}$$

The initial velocity of the 30-kg box of pineapples is zero: $\vec{v}_{20} = 0$.

$$(20)(9.9) + (30)(0) = (20 + 30)v$$

$$198 + 0 = 50v$$

$$v = \frac{198}{50} = 4.0 \text{ m/s}$$

The final speed of the boxes is $v = 4.0$ m/s.

24 TWO-DIMENSIONAL COLLISIONS

Conservation of Momentum

$$m_1 v_{10x} + m_2 v_{20x} = m_1 v_{1x} + m_2 v_{2x}$$
$$m_1 v_{10y} + m_2 v_{20y} = m_1 v_{1y} + m_2 v_{2y}$$

Components of Initial Velocity

$$v_{10x} = v_{10} \cos \theta_{10} \quad , \quad v_{20x} = v_{20} \cos \theta_{20}$$
$$v_{10y} = v_{10} \sin \theta_{10} \quad , \quad v_{20y} = v_{20} \sin \theta_{20}$$

Conservation of Kinetic Energy (Elastic Collision Only)

$$\frac{1}{2} m_1 v_{10}^2 + \frac{1}{2} m_2 v_{20}^2 = \frac{1}{2} m_1 v_1^2 + \frac{1}{2} m_2 v_2^2$$

Elastic Collision with Equal Masses

$$\theta_{1ref} + \theta_{2ref} = 90°$$

Final Speeds

$$v_1 = \sqrt{v_{1x}^2 + v_{1y}^2} \quad , \quad v_2 = \sqrt{v_{2x}^2 + v_{2y}^2}$$

Directions of the Final Velocities

$$\theta_1 = \tan^{-1}\left(\frac{v_{1y}}{v_{1x}}\right) \quad , \quad \theta_2 = \tan^{-1}\left(\frac{v_{2y}}{v_{2x}}\right)$$

Percent of Kinetic Energy Lost or Gained

$$KE_0 = \frac{1}{2} m_1 v_{10}^2 + \frac{1}{2} m_2 v_{20}^2 \quad , \quad KE = \frac{1}{2} m_1 v_1^2 + \frac{1}{2} m_2 v_2^2$$
$$\% \text{ change} = \frac{|KE - KE_0|}{KE_0} \times 100\%$$

Symbol	Name	Units
m_1	mass of object 1	kg
m_2	mass of object 2	kg
$\vec{v}_{10}$	initial velocity of object 1	m/s
$\vec{v}_{20}$	initial velocity of object 2	m/s
v_{10}	initial speed of object 1	m/s
v_{20}	initial speed of object 2	m/s
θ_{10}	initial direction of object 1	°
θ_{20}	initial direction of object 2	°
v_{10x}	initial x-component of velocity of object 1	m/s
v_{10y}	initial y-component of velocity of object 1	m/s
v_{20x}	initial x-component of velocity of object 2	m/s
v_{20y}	initial y-component of velocity of object 2	m/s
$\vec{v}_1$	final velocity of object 1	m/s
$\vec{v}_2$	final velocity of object 2	m/s
v_1	final speed of object 1	m/s
v_2	final speed of object 2	m/s
θ_1	final direction of object 1	°
θ_2	final direction of object 2	°
v_{1x}	final x-component of velocity of object 1	m/s
v_{1y}	final y-component of velocity of object 1	m/s
v_{2x}	final x-component of velocity of object 2	m/s
v_{2y}	final y-component of velocity of object 2	m/s

Example 147. A 300-kg bananamobile traveling 20 m/s to the north collides with a 200-kg bananamobile traveling 40 m/s to the east. The two bananamobiles stick together after the collision. Determine the speed of the bananamobiles just after the collision.

Solution. First setup a coordinate system. We choose $+x$ to point east and $+y$ to point north. With this choice, the 300-kg bananamobile's initial velocity ($\vec{v}_{10}$) will be along the unit vector $\hat{y}$, while the 200-kg bananamobile's initial velocity ($\vec{v}_{10}$) will be along the unit vector $\hat{x}$.

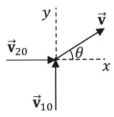

Identify the known quantities for object 1 (the 300-kg bananamobile) and object 2 (the 200-kg bananamobile).

- The mass of the 300-kg bananamobile is $m_1 = 300$ kg.
- The initial velocity of the 300-kg bananamobile is $\vec{v}_{10} = 20\hat{y}$ m/s. The initial speed and direction of object 1 are $v_{10} = 20$ m/s and $\theta_{10} = 90°$.
- The mass of the 200-kg bananamobile is $m_2 = 200$ kg.
- The initial velocity of the 200-kg bananamobile is $\vec{v}_{20} = 40\hat{x}$ m/s. The initial speed and direction of object 2 are $v_{20} = 40$ m/s and $\theta_{20} = 0°$.

Resolve the initial velocities into components.

$$v_{10x} = v_{10} \cos\theta_{10} = 20\cos 90° = 0 \quad , \quad v_{20x} = v_{20}\cos\theta_{20} = 40\cos 0° = 40 \text{ m/s}$$
$$v_{10y} = v_{10} \sin\theta_{10} = 20\sin 90° = 20 \text{ m/s} \quad , \quad v_{20y} = v_{20}\sin\theta_{20} = 40\sin 0° = 0$$

Since the two bananamobiles stick together, use the equations for a **perfectly inelastic collision**. The bananamobiles have the same final velocity.

$$m_1 v_{10x} + m_2 v_{20x} = (m_1 + m_2)v_x$$
$$300(0) + 200(40) = (300 + 200)v_x$$
$$m_1 v_{10y} + m_2 v_{20y} = (m_1 + m_2)v_y$$
$$300(20) + 200(0) = (300 + 200)v_y$$

Simplify each equation.

$$8000 = 500v_x$$
$$6000 = 500v_y$$

Divide both sides of each equation by 500.

$$v_x = \frac{8000}{500} = 16 \text{ m/s}$$
$$v_y = \frac{6000}{500} = 12 \text{ m/s}$$

Use the Pythagorean theorem to find the final speed from the components of the final velocity.

$$v = \sqrt{v_x^2 + v_y^2} = \sqrt{(16)^2 + (12)^2} = \sqrt{256 + 144} = \sqrt{400} = 20 \text{ m/s}$$

The final speed of the bananamobiles is $v = 20$ m/s.

Example 148. Two bananamobiles of equal mass collide at an intersection and stick together. One bananamobile is traveling 20 m/s to the south prior to the collision. After the collision, the two bananamobiles travel $10\sqrt{2}$ m/s to the southwest. Determine the initial speed and the direction of the initial velocity of the other bananamobile prior to the collision.

Solution. First setup a coordinate system. We choose $+x$ to point east and $+y$ to point north. With this choice, the first bananamobile's initial velocity ($\vec{v}_{10}$) will be along unit vector $-\hat{y}$, while the second bananamobile's initial velocity ($\vec{v}_{10}$) has unknown direction (we can reason that it points left, but it could be up to the left, directly left, or down to the left; once we do the math we will know which is correct).

Identify the known quantities for the two bananamobiles.

- The bananamobiles have the same (unkown) mass: $m_1 = m_2 = m$.
- The initial velocity of the first bananamobile is $\vec{v}_{10} = -20\hat{y}$ m/s. The initial speed and direction of object 1 are $v = 20$ m/s and $\theta = 270°$.
- The final speed and direction of the bananamobiles (after they stick together) are $v = 10\sqrt{2}$ m/s and $\theta = 225°$ (southwest lies in the middle of Quadrant III).

Resolve the initial velocity of object 1 into components.

$$v_{10x} = v_{10} \cos\theta_{10} = 20\cos 270° = 0 \quad , \quad v_{10y} = v_{10}\sin\theta_{10} = 20\sin 270° = -20 \text{ m/s}$$

Resolve the final velocity into components.

$$v_x = v\cos\theta = 10\sqrt{2}\cos 225° = -10 \text{ m/s} \quad , \quad v_y = v\sin\theta = 10\sqrt{2}\sin 225° = -10 \text{ m/s}$$

Since the two bananamobiles stick together, use the equations for a **perfectly inelastic collision**. The bananamobiles have the same final velocity.

$$m_1 v_{10x} + m_2 v_{20x} = (m_1 + m_2)v_x$$
$$m(0) + m v_{20x} = (m + m)(-10)$$
$$m_1 v_{10y} + m_2 v_{20y} = (m_1 + m_2)v_y$$
$$m(-20) + m v_{20y} = (m + m)(-10)$$

Recall that the bananamobiles have the same mass: $m_1 = m_2 = m$. Simplify each equation.

$$m v_{20x} = -20m$$
$$-20m + m v_{20y} = -20m$$

Divide both sides of each equation by m. The mass cancels out.
$$v_{20x} = -20 \text{ m/s}$$
$$-20 + v_{20y} = -20$$
Add 20 to both sides of the second equation.
$$v_{20y} = -20 + 20 = 0$$
The components of object 2's initial velocity are $v_{20x} = -20$ m/s and $v_{20y} = 0$. Use the Pythagorean theorem to find the initial speed of object 2 from the components of its final velocity.
$$v_{20} = \sqrt{v_{20x}^2 + v_{20y}^2} = \sqrt{(-20)^2 + (0)^2} = \sqrt{400} = 20 \text{ m/s}$$
The initial speed of object 2 is $v_{20} = 20$ m/s. Apply trig to find the direction of object 2's initial velocity.
$$\theta_{20} = \tan^{-1}\left(\frac{v_{20y}}{v_{20x}}\right) = \tan^{-1}\left(\frac{0}{-20}\right) = \tan^{-1}(0)$$
The reference angle is $0°$, but $0°$ isn't the answer. Since v_{20x} is negative and $v_{20y} = 0$, object 2's initial velocity is $\theta_{20} = 180°$ (which is to the west).

Example 149. A billiard ball traveling $6\sqrt{3}$ m/s collides with another billiard ball of equal mass that is initially at rest. After the collision, the first billiard ball travels $3\sqrt{3}$ m/s along a line that is deflected $60°$ relative to its original direction. The collision is <u>elastic</u>. Determine the magnitude and direction of the final velocity of the second billiard ball.

Solution. First setup a coordinate system. We choose $+x$ to point along the initial velocity of the first billiard ball and $+y$ to be perpendicular to that. With this choice, $\theta_{10} = 0°$ (along $+x$). Observe that $v_{20} = 0$ since the second billiard ball is initially at rest. Identify the known quantities for the two bananamobiles.

- The billiard balls have the same (unkown) mass: $m_1 = m_2 = m$.
- The initial velocity of the first billiard ball is $\vec{v}_{10} = 6\sqrt{3}\,\hat{x}$ m/s. The initial speed and direction of object 1 are $v = 6\sqrt{3}$ m/s and $\theta = 0°$.
- The final speed and direction of the first billiard ball are $v_{20} = 3\sqrt{3}$ m/s and $\theta_{20} = 60°$.

Resolve the initial velocities into components.
$$v_{10x} = v_{10}\cos\theta_{10} = 6\sqrt{3}\cos 0° = 6\sqrt{3} \text{ m/s} \quad , \quad v_{20x} = v_{20}\cos\theta_{20} = 0$$
$$v_{10y} = v_{10}\sin\theta_{10} = 6\sqrt{3}\sin 0° = 0 \quad , \quad v_{20y} = v_{20}\sin\theta_{20} = 0$$
Find the x- and y-components of the final velocity of object 1.
$$v_{1x} = v_1\cos\theta_1 = 3\sqrt{3}\cos 60° = \frac{3\sqrt{3}}{2}\frac{\text{m}}{\text{s}} \quad , \quad v_{1y} = v_1\sin\theta_1 = 3\sqrt{3}\sin 60° = \frac{9}{2} \text{ m/s}$$
Use the equations for a two-dimensional <u>elastic</u> collision with equal masses.

$$v_{10x} + v_{20x} = v_{1x} + v_{2x}$$
$$v_{10y} + v_{20y} = v_{1y} + v_{2y}$$
$$\theta_{1ref} + \theta_{2ref} = 90°$$

Plug numbers into these equations.

$$6\sqrt{3} + 0 = 6\sqrt{3} = \frac{3\sqrt{3}}{2} + v_{2x}$$

$$0 + 0 = 0 = \frac{9}{2} + v_{2y}$$

$$60° + \theta_{2ref} = 90°$$

Subtract $\frac{3\sqrt{3}}{2}$ from both sides of the first equation.

$$6\sqrt{3} - \frac{3\sqrt{3}}{2} = v_{2x}$$

$$\frac{12\sqrt{3}}{2} - \frac{3\sqrt{3}}{2} = v_{2x}$$

$$\frac{9\sqrt{3}}{2} \text{ m/s} = v_{2x}$$

Subtract $\frac{9}{2}$ from both sides of the second equation.

$$-\frac{9}{2} \text{ m/s} = v_{2y}$$

Subtract 60° from both sides of the third equation.
$$\theta_{2ref} = 90° - 60° = 30°$$

Use the components of the final velocity to determine the final speed and the direction of the final velocity for the second billiard ball.

$$v_2 = \sqrt{v_{2x}^2 + v_{2y}^2} = \sqrt{\left(\frac{9\sqrt{3}}{2}\right)^2 + \left(-\frac{9}{2}\right)^2} = \sqrt{\frac{(81)(3)}{4} + \frac{81}{4}} = \sqrt{\frac{324}{4}} = \sqrt{81} = 9.0 \text{ m/s}$$

$$\theta_2 = \tan^{-1}\left(\frac{v_{2y}}{v_{2x}}\right) = \tan^{-1}\left(\frac{-\frac{9}{2}}{\frac{9\sqrt{3}}{2}}\right) = \tan^{-1}\left(-\frac{9}{2} \times \frac{2}{9\sqrt{3}}\right) = \tan^{-1}\left(-\frac{1}{\sqrt{3}}\right) = \tan^{-1}\left(-\frac{\sqrt{3}}{3}\right)$$

To divide by a fraction, multiply by its **reciprocal**. Note that $\frac{1}{\sqrt{3}} = \frac{1}{\sqrt{3}}\frac{\sqrt{3}}{\sqrt{3}} = \frac{\sqrt{3}}{3}$. The reference angle is 30°, but θ_2 lies in Quadrant IV since v_{2x} is positive and v_{2y} is negative: $\theta_2 = \theta_{IV} = 360° - \theta_{ref} = 360° - 30° = 330°$. The final speed of the second billiard ball is $v_2 = 9.0$ m/s and its direction is $\theta_2 = 330°$.

25 ROCKET PROPULSION

Symbol	Name	SI Units
m_0	initial mass of the rocket (including all of the fuel)	kg
m_{f0}	mass of the initial fuel of the rocket	kg
m	mass of rocket plus the mass of any unburned fuel	kg
m_f	mass of the fuel remaining in the rocket	kg
m_p	the payload (the mass of the rocket without any fuel)	kg
r_p	the ratio of the payload (m_p) to the initial mass (m_0)	unitless
y	position coordinate	m
v_{y0}	initial y-component of the rocket's velocity	m/s
v_y	y-component of the rocket's velocity	m/s
$\vec{v}$	instantaneous velocity of the rocket relative to the inertial reference frame	m/s
$\vec{u}$	instantaneous velocity of the ejected gases relative to the rocket	m/s
$\vec{v} - \vec{u}$	instantaneous velocity of the ejected gases relative to the inertial reference frame	m/s
$\vec{P}_{tot}$	the total momentum of the system	kg·m/s
g	gravitational acceleration	m/s^2
a_y	y-component of acceleration	m/s^2
$\vec{F}$	force	N
$\vec{F}_t$	the rocket's thrust	N
R_b	the instantaneous rate at which the rocket burns fuel	kg/s
t	time	s
t_b	burn time (the time spent burning fuel)	s

Newton's Second Law (General Form)

$$\sum \vec{\mathbf{F}} = \frac{d\vec{\mathbf{p}}_{tot}}{dt}$$

Mass Ratios

$$r_p = \frac{m_p}{m_0} \quad , \quad m_0 = m_p + m_{f0} \quad , \quad m = m_p + m_f \quad , \quad \frac{m}{m_0} = \frac{m_p + m_f}{m_p + m_{f0}}$$

Burn Rate

$$R_b = \left|\frac{dm}{dt}\right|$$

Constant Burn Rate

$$m - m_0 = -R_b t_b$$

Thrust

$$\vec{\mathbf{F}}_t = -\vec{\mathbf{u}}\frac{dm}{dt}$$

One-dimensional Rocket Propulsion in an Approximately Uniform Gravitational Field with Constant Burn Rate

$$v_y = v_{y0} - g t_b + u \ln\left(\frac{m_0}{m}\right) \quad , \quad F_{thrust} - mg = ma_y$$

One-dimensional Rocket Propulsion in Zero-Gravity

$$v_y = v_{y0} + u \ln\left(\frac{m_0}{m}\right) \quad , \quad F_{thrust} = ma_y$$

Example 150. A 2,000-kg rocket (not including the fuel) in deep space – i.e. where gravity is negligible compared to the thrust of the rocket – has a payload of 25% and an initial speed of 200 m/s relative to the nearest (yet quite distant) star. Fuel is ejected with an exhaust speed of 1500 m/s relative to the rocket at a constant burn rate of 50 kg/s in order to accelerate the rocket to a final speed of 600 m/s.

(A) What percentage of the mass of the fuel is spent to accomplish this?

Solution. Begin by making a list of the given symbols along with their numerical values. We choose $+y$ to be in the direction of the rocket's motion.

- The payload is $m_p = 2000$ kg. This is the mass of the rocket without any fuel.
- The ratio of the payload (m_p) to the initial mass (m_0) is $r_p = 25\% = 0.25 = \frac{1}{4}$.
- The initial velocity of the rocket is $v_{y0} = 200$ m/s (relative to the nearest star).
- The exhaust speed is $u = 1500$ m/s (relative to the rocket).
- The final velocity of the rocket is $v_y = 600$ m/s (relative to the nearest star).
- The burn rate is $R_b = 50$ kg/s.

We need to find the final mass of the fuel (m_f) remaining in the rocket in order to figure out the answer to the question. We begin with the equation for the mass ratio.

$$r_p = \frac{m_p}{m_0}$$

Multiply both sides of the equation by the initial mass.

$$m_0 r_p = m_p$$

Divide both sides by the mass ratio to solve for the initial mass (m_0), including both the rocket and the initial fuel.

$$m_0 = \frac{m_p}{r_p} = \frac{2000}{0.25} = 8000 \text{ kg}$$

Subtract the payload to determine the initial mass of the fuel (m_{f0}).

$$m_{f0} = m_0 - m_p = 8000 - 2000 = 6000 \text{ kg}$$

Apply the zero-gravity equation of rocket propulsion.

$$v_y = v_{y0} + u \ln\left(\frac{m_0}{m}\right)$$

$$600 = 200 + 1500 \ln\left(\frac{8000}{m}\right)$$

Subtract 200 from both sides of the equation.

$$400 = 1500 \ln\left(\frac{8000}{m}\right)$$

Divide both sides of the equation by 1500.

$$0.267 = \ln\left(\frac{8000}{m}\right)$$

Exponentiate both sides of the equation.

$$e^{0.267} = e^{\ln\left(\frac{8000}{m}\right)}$$

Apply the rule $e^{\ln(x)} = x$.

$$e^{0.267} = \frac{8000}{m}$$

Multiply both sides by the final mass of the rocket including any unburned fuel.

$$m\, e^{0.267} = 8000$$

Divide both sides of the equation by $e^{0.267}$. Apply the rule $\frac{1}{e^{0.267}} = e^{-0.267}$.

$$m = \frac{8000}{e^{0.267}} = 8000 e^{-0.267} = 6125 \text{ kg}$$

Note that the numerical value of e (Euler's number) is approximately 2.718. Subtract the payload to determine the final mass of the fuel (m_f).

$$m_f = m - m_p = 6125 - 2000 = 4125 \text{ kg}$$

Compare the final mass of the fuel (m_f) to the initial mass of the fuel (m_{f0}) in order to determine the percentage of fuel spent.

$$\frac{m_{f0} - m_f}{m_{f0}} 100\% = \frac{6000 - 4125}{6000} 100\% = \frac{1875}{6000} 100\% = 31\%$$

The final answer is that 31% of the fuel is spent.

(B) How much time does this take?

Solution. Apply the equation for constant burn rate.

$$m - m_0 = -R_b t_b$$
$$6125 - 8000 = -50 t_b$$
$$-1875 = -50 t_b$$

Divide both sides of the equation by -50.

$$t_b = \frac{-1875}{-50} = 37.5 \text{ s}$$

The two minus signs make a plus sign. The burn time is $t_b = 37.5$ s.

Example 151. A 3,000-kg rocket (not including the fuel) has 6,000 kg of fuel and an initial speed of 100 m/s as it accelerates upward away from earth. With a constant burn rate, the rocket accelerates to 500 m/s in 12 s. The exhaust speed is 2000 m/s relative to the rocket. Assume that the rocket's initial altitude is low enough that any variation in gravitational acceleration from 9.81 m/s^2 may be neglected during this time interval.

(A) What percentage of the mass of the fuel is spent to accomplish this?

Solution. Begin by making a list of the given symbols along with their numerical values. We choose $+y$ to be upward.

- The payload is $m_p = 3000$ kg. This is the mass of the rocket without any fuel.
- The initial mass of the fuel is $m_{f0} = 6000$ kg.
- The initial velocity of the rocket is $v_{y0} = 100$ m/s (relative to the earth).
- The exhaust speed is $u = 2000$ m/s (relative to the rocket).
- The final velocity of the rocket is $v_y = 500$ m/s (relative to the earth).
- The earth's gravitational acceleration is $g = 9.81$ m/s^2. Note that this is **not** the rocket's acceleration.
- The burn time is $t_b = 12$ s.

We need to find the final mass of the fuel (m_f) remaining in the rocket in order to figure out the answer to the question. We begin by finding the initial mass of the rocket including the fuel.

$$m_0 = m_p + m_{f0} = 3000 + 6000 = 9000 \text{ kg}$$

Apply the uniform gravity equation of rocket propulsion.

$$v_y = v_{y0} - g t_b + u \ln\left(\frac{m_0}{m}\right)$$

$$500 = 100 - (9.81)(12) + 2000 \ln\left(\frac{9000}{m}\right)$$

Subtract 100 from both sides of the equation. Add $(9.81)(12)$ to both sides of the equation.

$$518 = 2000 \ln\left(\frac{9000}{m}\right)$$

Divide both sides of the equation by 2000.

$$0.259 = \ln\left(\frac{9000}{m}\right)$$

Exponentiate both sides of the equation.

$$e^{0.259} = e^{\ln\left(\frac{9000}{m}\right)}$$

Apply the rule $e^{\ln(x)} = x$.

$$e^{0.259} = \frac{9000}{m}$$

Multiply both sides by the final mass of the rocket including any unburned fuel.

$$m \, e^{0.259} = 9000$$

Divide both sides of the equation by $e^{0.259}$. Apply the rule $\frac{1}{e^{0.259}} = e^{-0.259}$.

$$m = \frac{9000}{e^{0.259}} = 9000e^{-0.259} = 6946 \text{ kg}$$

Subtract the payload to determine the final mass of the fuel (m_f).

$$m_f = m - m_p = 6946 - 3000 = 3946 \text{ kg}$$

Compare the final mass of the fuel (m_f) to the initial mass of the fuel (m_{f0}) in order to determine the percentage of fuel spent.

$$\frac{m_{f0} - m_f}{m_{f0}} 100\% = \frac{6000 - 3946}{6000} 100\% = \frac{2054}{6000} 100\% = 34\%$$

The final answer is that 34% of the fuel is spent.

(B) What is the burn rate?

Solution. Apply the equation for constant burn rate.

$$m - m_0 = -R_b t_b$$
$$6946 - 9000 = -R_b 12$$
$$-2054 = -12 R_b$$
$$R_b = \frac{-2054}{-12} = 171 \text{ kg/s}$$

The two minus signs make a plus sign. The burn rate is $R_b = 171$ kg/s.

(C) What is the thrust?

Solution. Use the equation for thrust.

$$F_t = -u\frac{dm}{dt}$$

Note that the burn rate is $R_b = -\frac{dm}{dt}$. The total mass of the rocket decreases in time as fuel is spent, so the minus sign is needed to make the burn rate positive.

$$F_t = -u(-R_b) = uR_b = (2000)(171) = 342{,}000 \text{ N} = 342 \text{ kN}$$

Recall that the prefix kilo (k) stands for 1000. The thrust is $F_t = 342$ kN.

(D) What is the initial acceleration?

Solution. Apply Newton's second law to the rocket.

$$F_{thrust} - m_0 g = m_0 a_{y0}$$

Divide both sides of the equation by the initial mass. Note that $\frac{m_0 g}{m_0} = g$.

$$a_{y0} = \frac{F_{thrust} - m_0 g}{m_0} = \frac{F_{thrust}}{m_0} - g = \frac{342{,}000}{9000} - 9.81 = 38 - 9.81 = 28 \text{ m/s}^2$$

The initial acceleration is $a_{y0} = 28$ m/s².

(E) What is the final acceleration?

Solution. Do the same as in part (D), but use the final mass instead of the initial mass.

$$a_{y0} = \frac{F_{thrust}}{m} - g = \frac{342{,}000}{6946} - 9.81 = 49 - 9.81 = 39 \text{ m/s}^2$$

The final acceleration is $a_y = 39$ m/s²

26 TECHNIQUES OF INTEGRATION

AND COORDINATE SYSTEMS

Example 152. Perform the following definite integral.

$$\int_{x=0}^{15} \left(\frac{x}{5} - 1\right)^4 dx$$

Solution. You can avoid having to multiply this out if you make the substitution $u = \frac{x}{5} - 1$. Take an implicit derivative of both sides of the equation.

- On the left-hand side: $\frac{d}{du}(u) = 1$. Multiply by du to get $1\, du = du$.

- On the right-hand side: $\frac{d}{dx}\left(\frac{x}{5} - 1\right) = \frac{1}{5}$. Multiply by dx to get $\frac{1}{5}dx$.

The implicit derivative of $u = \frac{x}{5} - 1$ is therefore $du = \frac{1}{5}dx$. Solve for dx to get $dx = 5du$.

$$u = \frac{x}{5} - 1 \quad , \quad dx = 5du$$

Plug the limits of x into the equation $u = \frac{x}{5} - 1$ in order to find the limits of u.

$$u(x = 0) = \frac{0}{5} - 1 = 0 - 1 = -1$$

$$u(x = 15) = \frac{15}{5} - 1 = 3 - 1 = 2$$

The new limits are from $u = -1$ to $u = 2$. Substitute $u = \frac{x}{5} - 1$, $dx = 5du$, and the new limits into the old integral over x to get a new integral over the variable u.

$$\int_{u=-1}^{2} u^4\, (5du) = \int_{u=-1}^{2} 5u^4\, du = \left[\frac{5u^{4+1}}{4+1}\right]_{u=-1}^{2} = \left[\frac{5u^5}{5}\right]_{u=-1}^{2} = [u^5]_{u=-1}^{2}$$

$$= (2)^5 - (-1)^5 = 32 - (-1) = 32 + 1 = 33$$

Note that $(-1)^5 = -1$ and $-(-1) = +1$. The final answer is 33.

Example 153. Perform the following definite integral (where the angles are in **radians**).

$$\int_{x=0}^{2\pi} \cos\left(\frac{\pi}{2} - \frac{x}{6}\right) dx$$

Solution. Make the substitution $u = \frac{\pi}{2} - \frac{x}{6}$ in order to simplify the argument of the cosine function. Take an implicit derivative of both sides of the equation.

- On the left-hand side: $\frac{d}{du}(u) = 1$. Multiply by du to get $1\, du = du$.

- On the right-hand side: $\frac{d}{dx}\left(\frac{\pi}{2} - \frac{x}{6}\right) = -\frac{x}{6}$. Multiply by dx to get $-\frac{1}{6}dx$.

The implicit derivative of $u = \frac{\pi}{2} - \frac{x}{6}$ is $du = -\frac{1}{6}dx$. Solve for dx to get $dx = -6du$.

$$u = \frac{\pi}{2} - \frac{x}{6} \quad , \quad dx = -6du$$

Plug the limits of x into the equation $u = \frac{\pi}{2} - \frac{x}{6}$ in order to find the limits of u.

$$u(x = 0) = \frac{\pi}{2} - \frac{0}{6} = \frac{\pi}{2} - 0 = \frac{\pi}{2}$$

$$u(x = 2\pi) = \frac{\pi}{2} - \frac{2\pi}{6} = \frac{3\pi}{6} - \frac{2\pi}{6} = \frac{3\pi - 2\pi}{6} = \frac{\pi}{6}$$

We subtracted the fractions by finding a **common denominator**. The new limits are from $u = \frac{\pi}{2}$ to $u = \frac{\pi}{6}$. Substitute $u = \frac{\pi}{2} - \frac{x}{6}$, $dx = -6du$, and the new limits into the old integral over x to get a new integral over the variable u.

$$\int_{u=\pi/2}^{\pi/6} \cos u \, (-6du) = \int_{u=\pi/2}^{\pi/6} (-6\cos u) \, du = [-6\sin u]_{u=-\pi/2}^{\pi/6}$$

$$= -6\sin\left(\frac{\pi}{6}\right) + 6\sin\left(\frac{\pi}{2}\right) = -6\left(\frac{1}{2}\right) + 6(1) = -3 + 6 = 3$$

Recall that the anti-derivative of the cosine function is the sine function: $\int \cos u \, du = \sin u$. Note that the angles are in **radians**. The final answer is 3.

Example 154. Perform the following definite integral.

$$\int_{x=0}^{\sqrt{3}} \frac{dx}{\sqrt{1+x^2}}$$

Solution. Make the substitution $x = \tan u$. The idea behind this is to apply the trig identity $1 + \tan^2 u = \sec^2 u$, such that $\sqrt{1+x^2}$ becomes $\sqrt{1 + \tan^2 u} = \sqrt{\sec^2 u} = \sec u$. The substitution $x = \tan u$ effectively removes the squareroot. Take an implicit derivative of both sides of the equation $x = \tan u$.

- On the left-hand side: $\frac{d}{dx}(x) = 1$. Multiply by dx to get $1 \, dx = dx$.

- On the right-hand side: $\frac{d}{du}(\tan u) = \sec^2 u$. Multiply by du to get $\sec^2 u \, du$.

The implicit derivative of $x = \tan u$ is therefore $dx = \sec^2 u \, du$.

$$x = \tan u \quad , \quad dx = \sec^2 u \, du$$

Solve for u by taking an inverse tangent of both sides of the equation: $u = \tan^{-1} x$. Plug the limits of x into the equation $u = \tan^{-1} x$ in order to find the limits of u.

$$u(x = 0) = \tan^{-1}(0) = 0°$$

$$u\left(x = \sqrt{3}\right) = \tan^{-1}\left(\sqrt{3}\right) = 60°$$

The new limits are from $u = 0°$ to $u = 60°$. Substitute $x = \tan u$, $dx = \sec^2 u \, du$, and the new limits into the old integral over x to get a new integral over the variable u.

$$\int_{u=0°}^{60°} \frac{\sec^2 u}{\sqrt{1 + \tan^2 u}} du$$

Apply the trig identity $1 + \tan^2 u = \sec^2 u$ to write $\sqrt{1 + \tan^2 u} = \sqrt{\sec^2 u} = \sec u$.

$$\int_{u=0°}^{60°} \frac{\sec^2 u}{\sqrt{1 + \tan^2 u}} du = \int_{u=0°}^{60°} \frac{\sec^2 u}{\sec u} du = \int_{u=0°}^{60°} \sec u \, du$$

Find the anti-derivative of secant at the end of Chapter 9.

$$\int_{u=0°}^{60°} \sec u \, du = [\ln|\sec u + \tan u|]_{u=0°}^{60°} = \ln|\sec 60° + \tan 60°| - \ln|\sec 0° + \tan 0°|$$

$$= \ln|2 + \sqrt{3}| - \ln|1 + 0| = \ln|2 + \sqrt{3}| - \ln|1| = \ln|2 + \sqrt{3}|$$

The last term vanishes because $\ln|1| = 0$. The final answer is $\ln|2 + \sqrt{3}|$.

Example 155. Perform the following definite integral.

$$\int_{x=0}^{2} \sqrt{4 - x^2} \, dx$$

Solution. Make the substitution $x = 2\sin u$. The idea behind this is to apply the trig identity $\sin^2 u + \cos^2 u = 1$, which can also be written $1 - \sin^2 u = \cos^2 u$. With the substitution $x = 2\sin u$, the squareroot $\sqrt{4 - x^2}$ will become $\sqrt{4 - 4\sin^2 u} = \sqrt{4\cos^2 u} = 2\cos u$. This substitution effectively removes the squareroot. Take an implicit derivative of both sides of the equation $x = 2\sin u$.

- On the left-hand side: $\frac{d}{dx}(x) = 1$. Multiply by dx to get $1 \, dx = dx$.

- On the right-hand side: $\frac{d}{du}(2\sin u) = 2\cos u$. Multiply by du to get $2\cos u \, du$.

The implicit derivative of $x = \tan u$ is therefore $dx = 2\cos u \, du$.

$$x = 2\sin u \quad , \quad dx = 2\cos u \, du$$

Solve for u by taking an inverse sine of both sides of the equation: $u = \sin^{-1}\left(\frac{x}{2}\right)$. Plug the limits of x into the equation $u = \sin^{-1}\left(\frac{x}{2}\right)$ in order to find the limits of u.

$$u(x = 0) = \sin^{-1}\left(\frac{0}{2}\right) = 0°$$

$$u(x = 2) = \sin^{-1}\left(\frac{2}{2}\right) = 90°$$

The new limits are from $u = 0°$ to $u = 90°$. Substitute $x = 2\sin u$, $dx = 2\cos u \, du$, and the new limits into the old integral over x to get a new integral over the variable u.

$$\int_{u=0°}^{60°} \sqrt{4 - (2\sin u)^2} \, (2\cos u \, du) = \int_{u=0°}^{90°} 2\cos u \sqrt{4 - 4\sin^2 u} \, du$$

Apply the trig identity $\sin^2 u + \cos^2 u = 1$, which can also be written as $1 - \sin^2 u = \cos^2 u$, to simplify the squareroot: $\sqrt{4 - 4\sin^2 u} = \sqrt{4\cos^2 u} = 2\cos u$.

$$\int_{u=0°}^{90°} 2\cos u\,(2\cos u)\,du = \int_{u=0°}^{90°} 4\cos^2 u\,du$$

Now apply the trig identity $\cos^2 u = \frac{1+\cos 2u}{2}$.

$$\int_{u=0°}^{90°} 4\cos^2 u\,du = \int_{u=0°}^{90°} 4\left(\frac{1+\cos 2u}{2}\right)du = \int_{u=0°}^{90°} (2 + 2\cos 2u)\,du$$

$$= \int_{u=0°}^{90°} 2\,du + \int_{u=0°}^{90°} 2\cos 2u\,du = [2u]_{u=0°}^{90°} + [\sin 2u]_{u=0°}^{90°} = [2u]_{u=0°}^{\pi/2} + [\sin 2u]_{u=0°}^{\pi/2}$$

$$= 2\left(\frac{\pi}{2}\right) - 2(0) + \sin\left(\frac{2\pi}{2}\right) - \sin(0) = \pi - 0 + \sin(\pi) - 0 = \pi - 0 + 0 - 0 = \pi$$

Note that $\frac{d}{du}(\sin 2u) = 2\cos 2u$ such that $\int 2\cos 2u\,du = \sin 2u$. We converted from degrees to radians, where $90° = \frac{\pi}{2}$ rad. Also note that $\sin(\pi) = \sin(180°) = 0$. The final answer for the integral is π.

Example 156. Perform the following double integral.

$$\int_{x=0}^{2} \int_{y=0}^{x^3} xy\,dx\,dy$$

Solution. Since the upper y-limit is a function of x, carry out the y-integration first.

$$\int_{x=0}^{2} \int_{y=0}^{x^3} xy\,dx\,dy = \int_{x=0}^{2} \left(\int_{y=0}^{x^3} xy\,dy\right)dx$$

When integrating over y, we treat x like any other constant. Therefore, we can factor x out of the y integral (but be careful not to pull x out of the x integral). We're applying the same concept as $\int cf(y)\,dy = c\int f(y)\,dy$.

$$\int_{x=0}^{2} \left(\int_{y=0}^{x^3} xy\,dy\right)dx = \int_{x=0}^{2} x\left(\int_{y=0}^{x^3} y\,dy\right)dx$$

To help make this clear, we will carry out the complete definite integral over y in parentheses before proceeding. Students who do this tend to make fewer mistakes when they're learning how to perform multiple integrals.

$$\int_{x=0}^{2} x\left(\int_{y=0}^{x^3} y\,dy\right)dx = \int_{x=0}^{2} x\left(\left[\frac{y^{1+1}}{1+1}\right]_{y=0}^{y=x^3}\right)dx = \int_{x=0}^{2} x\left(\left[\frac{y^2}{2}\right]_{y=0}^{y=x^3}\right)dx$$

$$= \int_{x=0}^{2} x\left(\frac{(x^3)^2}{2} - \frac{(0)^2}{2}\right)dx = \int_{x=0}^{2} x\left(\frac{x^6}{2}\right)dx = \int_{x=0}^{2} \frac{x^7}{2}\,dx$$

Now we have a single integral which we can integrate using the method from Chapter 2.

$$\int\limits_{x=0}^{2} \frac{x^7}{2} dx = \left[\frac{x^{7+1}}{2(7+1)}\right]_{x=0}^{x=2} = \left[\frac{x^8}{16}\right]_{x=0}^{x=2} = \frac{(2)^8}{16} - \frac{(0)^8}{16} = \frac{256}{16} - 0 = 16$$

The final answer to the double integral is 16.

Example 157. Perform the following double integral.

$$\int\limits_{x=0}^{5} \int\limits_{y=0}^{4} x^3 y^4 \, dxdy$$

Solution. Since all of the limits are constants, in this example we are free to integrate in any order. We will carry out the y-integration first.

$$\int\limits_{x=0}^{5} \int\limits_{y=0}^{4} x^3 y^4 \, dxdy = \int\limits_{x=0}^{5} \left(\int\limits_{y=0}^{4} x^3 y^4 \, dy\right) dx$$

When integrating over y, we treat x like any other constant. Therefore, we can factor x^3 out of the y integral (but be careful not to pull x^3 out of the x integral).

$$\int\limits_{x=0}^{5} \left(\int\limits_{y=0}^{4} x^3 y^4 \, dy\right) dx = \int\limits_{x=0}^{5} x^3 \left(\int\limits_{y=0}^{4} y^4 \, dy\right) dx = \int\limits_{x=0}^{5} x^3 \left(\left[\frac{y^{4+1}}{4+1}\right]_{y=0}^{y=4}\right) dx$$

$$= \int\limits_{x=0}^{5} x^3 \left(\left[\frac{y^5}{5}\right]_{y=0}^{y=4}\right) dx = \int\limits_{x=0}^{5} x^3 \left(\frac{4^5}{5} - \frac{0^5}{5}\right) dx = \int\limits_{x=0}^{5} \frac{4^5 x^3}{5} dx$$

Now we have a single integral which we can integrate using the method from Chapter 2.

$$\int\limits_{x=0}^{5} \frac{4^5 x^3}{5} dx = \left[\frac{4^5 x^{3+1}}{5(3+1)}\right]_{x=0}^{x=5} = \left[\frac{4^5 x^4}{20}\right]_{x=0}^{x=5} = \frac{4^5(5)^4}{20} - \frac{4^5(0)^4}{20} = \frac{4^5 5^4}{20} - 0 = 4^4 5^3 = 32,000$$

Note that $(4)(5)=20$, such that $\frac{4^5 5^4}{20} = \frac{4^5 5^4}{4^1 5^1} = 4^{5-1} 5^{4-1} = 4^4 5^3$. The final answer to the double integral is 32,000. It's the same as $4^4 5^3$.

Example 158. Perform the following double integral.

$$\int\limits_{x=0}^{4} \int\limits_{y=0}^{9} \sqrt{xy} \, dxdy$$

Solution. Since all of the limits are constants, in this example we are free to integrate in any order. We will carry out the y-integration first.

$$\int\limits_{x=0}^{4} \int\limits_{y=0}^{9} \sqrt{xy} \, dxdy = \int\limits_{x=0}^{4} \left(\int\limits_{y=0}^{9} \sqrt{xy} \, dy\right) dx$$

Apply the rule $\sqrt{xy} = x^{1/2} y^{1/2}$.

$$\int_{x=0}^{4}\left(\int_{y=0}^{9}\sqrt{xy}\,dy\right)dx = \int_{x=0}^{4}\left(\int_{y=0}^{9}x^{1/2}y^{1/2}\,dy\right)dx$$

When integrating over y, we treat x like any other constant. Therefore, we can factor $x^{1/2}$ out of the y integral (but be careful not to pull $x^{1/2}$ out of the x integral).

$$\int_{x=0}^{4}\left(\int_{y=0}^{9}x^{1/2}y^{1/2}\,dy\right)dx = \int_{x=0}^{4}x^{1/2}\left(\int_{y=0}^{9}y^{1/2}\,dy\right)dx = \int_{x=0}^{4}x^{1/2}\left(\left[\frac{y^{1/2+1}}{1/2+1}\right]_{y=0}^{y=9}\right)dx$$

$$= \int_{x=0}^{4}x^{1/2}\left(\left[\frac{y^{3/2}}{3/2}\right]_{y=0}^{y=9}\right)dx = \int_{x=0}^{4}x^{1/2}\left(\left[\frac{2y^{3/2}}{3}\right]_{y=0}^{y=9}\right)dx$$

To divide by a fraction, multiply by its **reciprocal**: $\frac{y^{3/2}}{3/2} = \frac{2y^{3/2}}{3}$.

$$\int_{x=0}^{4}x^{1/2}\left(\frac{2(9)^{3/2}}{3}-\frac{2(0)^{3/2}}{3}\right)dx = \int_{x=0}^{4}18x^{1/2}\,dx$$

Note that $9^{3/2} = \left(\sqrt{9}\right)^{3} = 27$ and $\frac{2}{3}(27) = 18$. Now we have a single integral which we can integrate using the method from Chapter 2.

$$\int_{x=0}^{4}18x^{1/2}\,dx = \left[\frac{18x^{1/2+1}}{1/2+1}\right]_{x=0}^{x=4} = \left[\frac{18x^{3/2}}{3/2}\right]_{x=0}^{x=4} = \left[12x^{3/2}\right]_{x=0}^{x=4}$$

To divide by a fraction, multiply by its **reciprocal**: $\frac{18x^{3/2}}{3/2} = \left(18x^{3/2}\right)\frac{2}{3} = 12x^{3/2}$.

$$\left[12x^{3/2}\right]_{x=0}^{x=4} = 12(4)^{3/2} - 12(0)^{3/2} = 12(8) - 0 = 96$$

Note that $4^{3/2} = \left(\sqrt{4}\right)^{3} = 8$. The final answer to the double integral is 96.

Example 159. Perform the following double integral.

$$\int_{x=0}^{3}\int_{y=0}^{x}6(x-y)\,dxdy$$

Solution. First distribute the 6 according to $6(x-y) = 6x - 6y$ and split the integral into two separate integrals using the rule $\int(y_1+y_2)\,dx = \int y_1\,dx + \int y_2\,dx$.

$$\int_{x=0}^{3}\int_{y=0}^{x}6(x-y)\,dxdy = \int_{x=0}^{3}\int_{y=0}^{x}6x\,dxdy - \int_{x=0}^{3}\int_{y=0}^{x}6y\,dxdy$$

Since the upper y-limit is a function of x, carry out the y-integration first.

$$\int_{x=0}^{3}\int_{y=0}^{x}6x\,dxdy - \int_{x=0}^{3}\int_{y=0}^{x}6y\,dxdy = \int_{x=0}^{3}\left(\int_{y=0}^{x}6x\,dy\right)dx - \int_{x=0}^{3}\left(\int_{y=0}^{x}6y\,dy\right)dx$$

When integrating over y, we treat x like any other constant. Therefore, we can factor x out of the y integral (but be careful not to pull x out of the x integral).

$$\int_{x=0}^{3}\left(\int_{y=0}^{x} 6x\,dy\right)dx - \int_{x=0}^{3}\left(\int_{y=0}^{x} 6y\,dy\right)dx = \int_{x=0}^{3} 6x\left(\int_{y=0}^{x} dy\right)dx - \int_{x=0}^{3} 6\left(\int_{y=0}^{x} y\,dy\right)dx$$

$$= \int_{x=0}^{3} 6x\left([y]_{y=0}^{y=x}\right)dx - \int_{x=0}^{3} 6\left(\left[\frac{y^2}{2}\right]_{y=0}^{y=x}\right)dx = \int_{x=0}^{3} 6x(x-0)\,dx - \int_{x=0}^{3} 6\left(\frac{x^2}{2} - \frac{0^2}{2}\right)dx$$

$$= \int_{x=0}^{3} 6x(x)\,dx - \int_{x=0}^{3} 6\left(\frac{x^2}{2}\right)dx = \int_{x=0}^{3} 6x^2\,dx - \int_{x=0}^{3} 3x^2\,dx$$

Note that $\int dy = y$ and $\int y\,dy = \frac{y^2}{2}$. We can combine the last two integrals the same way that we separated the original integral back in the beginning of our solution.

$$\int_{x=0}^{3} 6x^2\,dx - \int_{x=0}^{3} 3x^2\,dx = \int_{x=0}^{3}(6x^2 - 3x^2)\,dx = \int_{x=0}^{3} 3x^2\,dx$$

Now we have a single integral which we can integrate using the method from Chapter 2.

$$\int_{x=0}^{3} 3x^2\,dx = \left[\frac{3x^{2+1}}{2+1}\right]_{x=0}^{x=3} = \left[\frac{3x^3}{3}\right]_{x=0}^{x=3} = [x^3]_{x=0}^{x=3} = (3)^3 - (0)^3 = 27 - 0 = 27$$

The final answer to the double integral is 27.

Example 160. Perform the following triple integral.

$$\int_{x=0}^{2}\int_{y=0}^{x^2}\int_{z=0}^{y} 27z^2\,dx\,dy\,dz$$

Solution. First integrate over z.

$$\int_{x=0}^{2}\int_{y=0}^{x^2} 27\left(\int_{z=0}^{y} z^2\,dz\right)dy\,dx = \int_{x=0}^{2}\int_{y=0}^{x^2} 27\left(\left[\frac{z^{2+1}}{2+1}\right]_{z=0}^{z=y}\right)dy\,dx$$

$$= \int_{x=0}^{2}\int_{y=0}^{x^2} 27\left(\left[\frac{z^3}{3}\right]_{z=0}^{z=y}\right)dy\,dx = \int_{x=0}^{2}\int_{y=0}^{x^2} 27\left(\frac{y^3}{3} - \frac{0}{3}\right)dy\,dx = \int_{x=0}^{2}\int_{y=0}^{x^2} 9y^3\,dy\,dx$$

Next integrate over y since the y-integration has a variable limit.

$$\int_{x=0}^{2}\int_{y=0}^{x^2} 9y^3\,dy\,dx = \int_{x=0}^{2}\left(\int_{y=0}^{x^2} 9y^3\,dy\right)dx = \int_{x=0}^{2}\left(\left[\frac{9y^{3+1}}{3+1}\right]_{y=0}^{y=x^2}\right)dx$$

$$= \int_{x=0}^{2}\left(\left[\frac{9y^4}{4}\right]_{y=0}^{y=x^2}\right)dx = \int_{x=0}^{2}\left(\frac{9(x^2)^4}{4} - \frac{9(0)^4}{4}\right)dx = \int_{x=0}^{2}\frac{9x^8}{4}\,dx$$

Now integrate over x.

$$\int_{x=0}^{2} \frac{9x^8}{4} dx = \left[\frac{9x^{8+1}}{4(8+1)}\right]_{x=0}^{x=2} = \left[\frac{9x^9}{4(9)}\right]_{x=0}^{x=2} = \left[\frac{x^9}{4}\right]_{x=0}^{x=2} = \frac{(2)^9}{4} - \frac{(0)^9}{4} = \frac{512}{4} - 0 = 128$$

The final answer to the triple integral is 128.

Example 161. Perform the following triple integral.

$$\int_{x=0}^{z} \int_{y=0}^{\sqrt{x}} \int_{z=0}^{9} 14xy^2 \, dx \, dy \, dz$$

Solution. First integrate over y. Apply the rule $\sqrt{x} = x^{1/2}$.

$$\int_{z=0}^{9} \int_{x=0}^{z} 14x \left(\int_{y=0}^{x^{1/2}} y^2 \, dy \right) dx \, dz = \int_{z=0}^{9} \int_{x=0}^{z} 14x \left(\left[\frac{y^{2+1}}{2+1}\right]_{y=0}^{y=x^{1/2}} \right) dx \, dz$$

$$= \int_{z=0}^{9} \int_{x=0}^{z} 14x \left(\left[\frac{y^3}{3}\right]_{y=0}^{y=x^{1/2}} \right) dx \, dz = \int_{z=0}^{9} \int_{x=0}^{z} 14x \left(\frac{x^{3/2}}{3} - \frac{0^3}{3} \right) dx \, dz$$

$$= \int_{z=0}^{9} \int_{x=0}^{z} 14x \left(\frac{x^{3/2}}{3} \right) dx \, dz = \int_{z=0}^{9} \int_{x=0}^{z} \frac{14x^{5/2}}{3} dx \, dz$$

Note that $\left(x^{1/2}\right)^3 = x^{3/2}$ according to the rule $(x^m)^n = x^{mn}$ and $xx^{3/2} = x^1 x^{3/2} = x^{5/2}$ according to the rule $x^m x^n = x^{m+n}$. Add fractions using a **common denominator**: $1 + \frac{3}{2} = \frac{2}{2} + \frac{3}{2} = \frac{2+3}{2} = \frac{5}{2}$. Next integrate over x since the x-integration has a variable limit.

$$\int_{z=0}^{9} \int_{x=0}^{z} \frac{14x^{5/2}}{3} dx \, dz = \int_{z=0}^{9} \left(\int_{x=0}^{z} \frac{14x^{5/2}}{3} dx \right) dz = \int_{z=0}^{9} \left(\left[\frac{14x^{5/2+1}}{3(5/2+1)}\right]_{x=0}^{x=z} \right) dz$$

Add fractions using a **common denominator**: $1 + \frac{5}{2} = \frac{2}{2} + \frac{5}{2} = \frac{2+5}{2} = \frac{7}{2}$.

$$\int_{z=0}^{9} \left(\left[\frac{14x^{5/2+1}}{3(5/2+1)}\right]_{x=0}^{x=z} \right) dz = \int_{z=0}^{9} \left(\left[\frac{14x^{7/2}}{3(7/2)}\right]_{x=0}^{x=z} \right) dz$$

To divide by a fraction, multiply by its **reciprocal**.

$$\int_{z=0}^{9} \left(\left[\frac{14x^{7/2}}{3(7/2)}\right]_{x=0}^{x=z} \right) dz = \int_{z=0}^{9} \left(\left[\frac{14x^{7/2}}{3} \frac{2}{7}\right]_{x=0}^{x=z} \right) dz = \int_{z=0}^{9} \left(\left[\frac{4x^{7/2}}{3}\right]_{x=0}^{x=z} \right) dz$$

Note that $\left(\frac{14}{3}\right)\left(\frac{2}{7}\right) = \frac{28}{21} = \frac{28 \div 7}{21 \div 7} = \frac{4}{3}$.

$$\int_{z=0}^{9} \left(\left[\frac{4x^{7/2}}{3}\right]_{x=0}^{x=z} \right) dz = \int_{z=0}^{9} \left(\frac{4z^{7/2}}{3} - \frac{4(0)^{7/2}}{3} \right) dz = \int_{z=0}^{9} \frac{4z^{7/2}}{3} dz$$

Now integrate over z.

198

$$\int_{z=0}^{9} \frac{4z^{7/2}}{3} dz = \left[\frac{4z^{7/2+1}}{3(7/2+1)} \right]_{z=0}^{z=9}$$

Add fractions using a **common denominator**: $1 + \frac{7}{2} = \frac{2}{2} + \frac{7}{2} = \frac{2+7}{2} = \frac{9}{2}$.

$$\left[\frac{4z^{7/2+1}}{3(7/2+1)} \right]_{z=0}^{z=9} = \left[\frac{4z^{9/2}}{3(9/2)} \right]_{z=0}^{z=9}$$

To divide by a fraction, multiply by its **reciprocal**.

$$\left[\frac{4z^{9/2}}{3(9/2)} \right]_{z=0}^{z=9} = \left[\frac{4z^{9/2}}{3} \frac{2}{9} \right]_{z=0}^{z=9} = \left[\frac{8z^{9/2}}{27} \right]_{z=0}^{z=9} = \frac{8(9)^{9/2}}{27} - \frac{8(0)^{9/2}}{27} = \frac{8(9)^{9/2}}{27} - 0 = \frac{8(9)^{9/2}}{27}$$

Note that $9^{9/2} = \left(\sqrt{9}\right)^9 = 3^9 = 19{,}683$. Alternatively, enter 9^(9/2) or 9^4.5 on your calculator to see that it equals 19,683 (in the former case, be sure to use parentheses correctly).

$$\frac{8(9)^{9/2}}{27} = \frac{8(19{,}683)}{27} = 5832$$

The final answer to the triple integral is 5832.

Differential Arc Length		
$ds = dx$ (along x)	$ds = dy$ (along y)	$ds = Rd\theta$ (circular arc)
Differential Area Element		
$dA = dxdy$ (polygon)	$dA = rdrd\theta$ (pie slice)	$dA = R^2 \sin\theta \, d\theta d\varphi$ (spherical surface)
Differential Volume Element		
$dV = dxdydz$ (bounded by flat sides)	$dV = r_c dr_c d\theta dz$ (cylinder or cone)	$dV = r^2 \sin\theta \, dr d\theta d\varphi$ (spherical)

Cartesian Coordinates	2D Polar Coordinates
	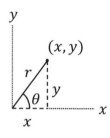 $$x = r\cos\theta \quad , \quad r = \sqrt{x^2 + y^2}$$ $$y = r\sin\theta \quad , \quad \theta = \tan^{-1}\left(\frac{y}{x}\right)$$
Cylindrical Coordinates	**Spherical Coordinates**
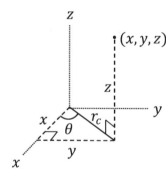 $$x = r_c\cos\theta \quad , \quad r_c = \sqrt{x^2 + y^2}$$ $$y = r_c\sin\theta \quad , \quad \theta = \tan^{-1}\left(\frac{y}{x}\right)$$	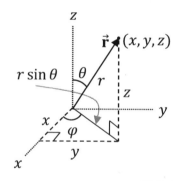 $$x = r\sin\theta\cos\varphi \quad , \quad r = \sqrt{x^2 + y^2 + z^2}$$ $$y = r\sin\theta\sin\varphi \quad , \quad \theta = \cos^{-1}\left(\frac{z}{r}\right)$$ $$z = r\cos\theta \quad , \quad \varphi = \tan^{-1}\left(\frac{y}{x}\right)$$

Example 162. Find the area of the triangle illustrated below using a double integral.

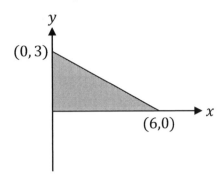

Solution. For a triangle, use Cartesian coordinates: Express the differential area element as $dA = dxdy$. This requires integrating over both x and y. We **don't** want to let x vary from 0 to 6 and let y vary from 0 to 3 because that would give us a rectangle instead of a triangle. If we let x vary from 0 to 6, observe that for a given value of x, y varies from 0 to the hypotenuse of the triangle, which is less than 3 (except when x reaches its lower limit). See the diagram below.

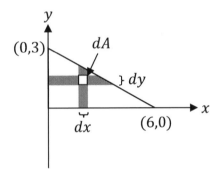

To find the proper upper limit for y, we need the equation of the line that serves as the triangle's hypotenuse. That line has a slope equal to $\frac{y_2 - y_1}{x_2 - x_1} = \frac{0-3}{6-0} = -\frac{3}{6} = -\frac{1}{2}$ and a y-intercept of 3. Since the general equation for a line is $y = mx + b$, the equation for this line is $y = -\frac{1}{2}x + 3$. For a given value of x, y will vary from 0 to $y = -\frac{1}{2}x + 3$, where the upper limit came from the equation for the line (of the hypotenuse). See the vertical gray band in the previous diagram. Now we have the integration limits.

$$A = \int_{x=0}^{6} \int_{y=0}^{-\frac{x}{2}+3} dy\, dx$$

We must perform the y-integration first because it has x in its upper limit.

$$A = \int_{x=0}^{6} \left(\int_{y=0}^{-\frac{x}{2}+3} dy \right) dx = \int_{x=0}^{6} \left([y]_{y=0}^{y=-\frac{x}{2}+3} \right) dx = \int_{x=0}^{6} \left(-\frac{x}{2} + 3 - 0 \right) dx = \int_{x=0}^{6} \left(-\frac{x}{2} + 3 \right) dx$$

$$= \left[-\frac{x^2}{4} + 3x \right]_{x=0}^{x=6} = -\frac{(6)^2}{4} + 3(6) - \left[-\frac{(0)^2}{4} + 3(0) \right] = -\frac{36}{4} + 18 = -9 + 18 = 9$$

The final answer to the double integral is $A = 9$. Of course, we don't need calculus to find the area of a triangle. We could just use the formula one-half base times height (Chapter 5): $A = \frac{1}{2}bh = \frac{1}{2}(6)(3) = 9$. However, there are some integrals that can only be done with calculus. (On a similar note, while area can be calculated as a single integral, there are some physics integrals that can only be done as double or triple integrals, so multi-integration is a necessary skill.) We will see examples of integrals of this form in physics problems in Chapters 27 and 32.

Example 163. A thick circular ring has inner radius R_1 and outer radius R_2. Find the area of this ring using a double integral.

Solution. For a solid circular ring, use $dA = r\,dr\,d\theta$. This requires integrating over both r and θ. Unlike in the previous example, r and θ each have constant limits:

$$R_1 \leq r \leq R_2$$
$$0 \leq \theta \leq 2\pi$$

For any value of θ, the variable r ranges between R_1 and R_2. Contrast this with the previous example where the upper limit of y depended on the value of x. Substitute $dA = r\,dr\,d\theta$ into the area integral.

$$A = \int dA = \int_{r=R_1}^{R_2} \int_{\theta=0}^{2\pi} r\,dr\,d\theta$$

Since all of the limits are constant, we can do these integrals in any order.

$$A = \int_{r=R_1}^{R_2} \left(\int_{\theta=0}^{2\pi} r\,d\theta \right) dr$$

When integrating over θ, treat the independent variable r as a constant. This means that you can pull r out of the θ integral (but be careful not to pull r out of the r integral).

$$A = \int_{r=R_1}^{R_2} r \left(\int_{\theta=0}^{2\pi} d\theta \right) dr = \int_{r=R_1}^{R_2} r[\theta]_{\theta=0}^{2\pi}\,dr = \int_{r=R_1}^{R_2} r(2\pi - 0)\,dr = \int_{r=R_1}^{R_2} 2\pi r\,dr$$

$$A = \int_{r=R_1}^{R_2} 2\pi r\,dr = 2\pi \int_{r=R_1}^{R_2} r\,dr = 2\pi \left[\frac{r^2}{2} \right]_{r=R_1}^{R_2} = 2\pi \left(\frac{R_2^2}{2} - \frac{R_1^2}{2} \right) = \pi(R_2^2 - R_1^2)$$

The final answer to the double integral is $A = \pi(R_2^2 - R_1^2)$. It's the same as the area of the outer circle (πR_2^2) minus the area of the inner circle (πR_1^2).

27 CENTER OF MASS

x-Coordinate of the Center of Mass
$$x_{cm} = \frac{m_1 x_1 + m_2 x_2 + \cdots + m_N x_N}{m_1 + m_2 + \cdots + m_N}$$
y-Coordinate of the Center of Mass
$$y_{cm} = \frac{m_1 y_1 + m_2 y_2 + \cdots + m_N y_N}{m_1 + m_2 + \cdots + m_N}$$

Symbol	Name	SI Units
m_i	mass of object i	kg
x_i	x-coordinate of object i	m
y_i	y-coordinate of object i	m
N	number of objects in the system	unitless
x_{cm}	x-coordinate of the center of mass	m
y_{cm}	y-coordinate of the center of mass	m
(x_{cm}, y_{cm})	coordinates of the center of mass	m

Example 164. A 150-g banana lies 250 cm from a 350-g bunch of bananas. Where is the center of mass of the system?

Solution. We setup our coordinate system with the origin on the 150-g banana and with $+x$ pointing toward the bunch of bananas. We call object 1 the banana and object 2 the bunch of bananas. Identify the information given in the problem.

- The banana's mass is $m_1 = 150$ g.
- The banana lies on our origin such that $x_1 = 0$.
- The bunch of bananas has a mass equal to $m_2 = 350$ g.
- The bunch of bananas lies at $x_2 = 250$ cm.

Note that in this problem it's not necessary to convert mass to kilograms or to convert length to meters, so long as we're consistent (don't mix and match grams with kilograms, or meters with centimeters). The grams will cancel out.

Use the equation for center of mass with $N = 2$ (for two objects).
$$x_{cm} = \frac{m_1 x_1 + m_2 x_2}{m_1 + m_2} = \frac{(150)(0) + (350)(250)}{150 + 350} = \frac{0 + 87{,}500}{500} = \frac{87{,}500}{500} = 175 \text{ cm}$$
The x-coordinate of the center of mass is $x_{cm} = 175$ cm. This means that the center of mass lies 175 cm from the 150-g banana (or 75 cm from the bunch of bananas). The y-coordinate of the center of mass is $y_{cm} = 0$ since both objects lie on the x-axis (meaing that $y_1 = y_2 = 0$). The coordinates of the center of mass are therefore $(x_{cm}, y_{cm}) = (175 \text{ cm}, 0)$.

Example 165. Three bananas have the following masses and coordinates. Find the location of the center of mass of the system.

- Object 1 has a mass of 200 g. Its center of mass lies at (7.0 m, 1.0 m).
- Object 2 has a mass of 500 g. Its center of mass lies at (6.0 m, 3.0 m).
- Object 3 has a mass of 800 g. Its center of mass lies at (2.0 m, –4.0 m).

Solution. Identify the information given in the problem.

- The masses are $m_1 = 200$ g, $m_2 = 500$ g, and $m_3 = 800$ g.
- The x-coordinates are $x_1 = 7.0$ m, $x_2 = 6.0$ m, and $x_3 = 2.0$ m.
- The y-coordinates are $y_1 = 1.0$ m, $y_2 = 3.0$ m, and $y_3 = -4.0$ m.

Use the equations for center of mass with $N = 3$ (for 3 objects).
$$x_{cm} = \frac{m_1 x_1 + m_2 x_2 + m_3 x_3}{m_1 + m_2 + m_3} = \frac{(200)(7) + 500(6) + 800(2)}{200 + 500 + 800} = \frac{6000}{1500} = 4.0 \text{ m}$$
$$y_{cm} = \frac{m_1 y_1 + m_2 y_2 + m_3 y_3}{m_1 + m_2 + m_3} = \frac{(200)(1) + 500(3) + 800(-4)}{200 + 500 + 800} = \frac{-1500}{1500} = -1.0 \text{ m}$$
The coordinates of the center of mass are $x_{cm} = 4.0$ m and $y_{cm} = -1.0$ m, which can also be expressed in the form $(x_{cm}, y_{cm}) = (4.0 \text{ m}, -1.0 \text{ m})$.

Example 166. The T-shaped object shown below consists of a 20.0-cm long handle and a 4.0-cm wide end. Each piece has uniform density (but the densities of the two pieces differ). The mass of the handle is 6.0 kg, while the mass of the end is 18.0 kg. Where could a fulcrum be placed below the object such that the object would be balanced?

4.0 cm

20.0 cm

Solution. We setup our coordinate system with the origin at the left end with $+x$ pointing to the right. Split the T-shaped object into two pieces: Object 1 is the 20.0-cm long handle and object 2 is the 4.0-cm wide end. Identify the information given in the problem.

- Object 1 (the 20-cm long handle) has a mass equal to $m_1 = 6.0$ kg.
- The x-coordinate of the center of object 1 is $x_1 = 10.0$ cm. Note that the handle's center lies at the midpoint of the handle and recall that we put our origin at the left end of the handle: $x_1 = \frac{L}{2} = \frac{20}{2} = 10.0$ cm.
- Object 2 (the 4.0-cm wide end) has a mass equal to $m_2 = 18.0$ kg.
- The x-coordinate of the center of object 1 is $x_2 = 22.0$ cm. Note that the center of object 2 is 2.0 cm from the point where the two objects meet, which is 20.0 cm to the right of the origin: $x_2 = L + \frac{W}{2} = 20 + \frac{4}{2} = 20 + 2 = 22.0$ cm.

Use the equation for center of mass with $N = 2$ (for two objects).
$$x_{cm} = \frac{m_1 x_1 + m_2 x_2}{m_1 + m_2} = \frac{(6)(10) + (18)(22)}{6 + 18} = \frac{60 + 396}{24} = \frac{456}{24} = 19.0 \text{ cm}$$
The x-coordinate of the center of mass is $x_{cm} = 19.0$ cm. The center of mass lies 1.0 cm to the left of the point where the handle meets the end (since $20.0 - 19.0 = 1.0$ cm).

Example 167. Find the center of mass of the system illustrated below <u>in gray</u>, where each gray square has the same uniform density and an edge length of 20 m.

Solution. First find the center of each gray square, knowing that the dimensions of each gray square are 20 m × 20 m. The centers of the left squares have $x = \frac{L}{2} = 10$ m and the centers of the right squares have $x = L + \frac{L}{2} = 20 + \frac{20}{2} = 30$ m. Similarly, the centers of the bottom, middle, and top squares have $y = 10$ m, 30 m, and 50 m, respectively.

- The bottom left gray square has a center at $(x_1, y_1) = (10 \text{ m}, 10 \text{ m})$.
- The bottom right gray square has a center at $(x_2, y_2) = (30 \text{ m}, 10 \text{ m})$.
- The middle right gray square has a center at $(x_3, y_3) = (30 \text{ m}, 30 \text{ m})$.
- The top right gray square has a center at $(x_4, y_4) = (30 \text{ m}, 50 \text{ m})$.
- The top left gray square has a center at $(x_5, y_5) = (10 \text{ m}, 50 \text{ m})$.

Make a list of the known symbols.

- All of the masses are identical: Call each mass m_s.
- The x-coordinates are $x_1 = 10$ m, $x_2 = 30$ m, $x_3 = 30$ m, $x_4 = 30$ m, and $x_5 = 10$ m.
- The y-coordinates are $y_1 = 10$ m, $y_2 = 10$ m, $y_3 = 30$ m, $y_4 = 50$ m, and $y_5 = 50$ m.

Use the equations for center of mass with $N = 5$ (for 5 squares).

$$x_{cm} = \frac{m_s x_1 + m_s x_2 + m_s x_3 + m_s x_4 + m_s x_5}{m_s + m_s + m_s + m_s + m_s} = \frac{10m_s + 30m_s + 30m_s + 30m_s + 10m_s}{5m_s}$$

$$= \frac{110 m_s}{5 m_s} = \frac{110}{5} = 22 \text{ m}$$

$$y_{cm} = \frac{m_s y_1 + m_s y_2 + m_s y_3 + m_s y_4 + m_s y_5}{m_s + m_s + m_s + m_s + m_s} = \frac{10m_s + 10m_s + 30m_s + 50m_s + 50m_s}{5m_s}$$

$$= \frac{150 m_s}{5 m_s} = \frac{150}{5} = 30 \text{ m}$$

Note that the mass m_s cancels out. The coordinates of the center of mass are $x_{cm} = 22$ m and $y_{cm} = 30$ m, which can also be expressed in the form $(x_{cm}, y_{cm}) = (22 \text{ m}, 30 \text{ m})$.

Example 168. As illustrated below, a circular hole with a radius of 2.0 m is cut out of an otherwise uniform circular solid disc with a radius of 8.0 m. Where is the center of mass of this object?

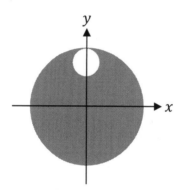

Solution. One way to solve this problem is to visualize the complete circle as the sum of the missing piece plus the shape with the hole cut out of it, as illustrated below.

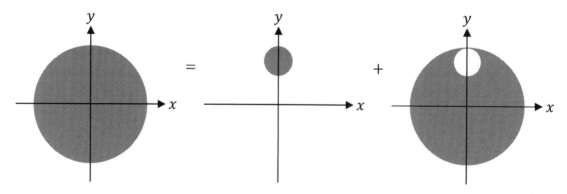

Write an equation for the center of mass of the large circle. (We'll only do this for y, since all three shapes obviously have $x_{cm} = 0$, as each one is centered about the y-axis.)

$$y_{cm} = \frac{m_1 y_1 + m_2 y_2}{m_1 + m_2}$$

Find the center of each shape.
- The center of the large circle has $y_{cm} = 0$ because it is centered about the origin.
- The center of the small circle (object 1) has $y_1 = 6.0$ m because the small circle's center lies 2.0 m below the top of the large circle: $y_1 = R_L - R_S = 8 - 2 = 6.0$ m.
- The shape with the hole cut out of it (object 2) has its center of mass in an unknown location. We're solving for y_2 in this problem.

Plug in the values for y_{cm} and y_1 into the equation for the center of mass of the large circle.

$$y_{cm} = \frac{m_1 y_1 + m_2 y_2}{m_1 + m_2}$$

$$0 = \frac{6m_1 + m_2 y_2}{m_1 + m_2}$$

Multiply both sides by $m_1 + m_2$.

$$0(m_1 + m_2) = m_1 6 + m_2 y_2$$

Zero times anything equals zero. Therefore, $0(m_1 + m_2) = 0$.

$$0 = m_1 6 + m_2 y_2$$

The mass of each object is proportional to its area: $m = \sigma A$, where the proportionality constant is the lowercase Greek symbol sigma (σ). Recall that the formula for the area of a circle is $A = \pi R^2$ (see Chapter 5).

- The large circle has mass $m_L = \sigma A_L = \sigma \pi R_L^2 = \sigma \pi (8)^2 = 64\sigma\pi$.
- The small circle has mass $m_1 = \sigma A_1 = \sigma \pi R_1^2 = \sigma \pi (2)^2 = 4\sigma\pi$.
- The shape with the hole cut out has mass $m_2 = m_L - m_1 = 64\sigma\pi - 4\sigma\pi = 60\sigma\pi$.

Plug all of these values into the equation above involving y_1 and y_2.

$$0 = m_1 6 + m_2 y_2$$

$$0 = (4\sigma\pi)(6) + 60\sigma\pi y_2$$

Divide both sides of the equation by $\sigma\pi$. The constants σ and π will cancel out.

$$0 = 24 + 60 y_2$$

Subtract 24 from both sides of the equation.

$$-24 = 60 y_2$$

Divide both sides of the equation by 60.

$$y_2 = -\frac{24}{60} = -\frac{2}{5} \text{ m} = -0.40 \text{ m}$$

The y-coordinate of the center of mass of the given shape is $y_2 = -\frac{2}{5}$ m. It's the same as $y_2 = -0.40$ m. The coordinates of the center of mass can also be expressed in the form $(x_{cm}, y_{cm}) = \left(0, -\frac{2}{5} \text{ m}\right) = (0, -0.40 \text{ m})$.

Coordinates of the Center of Mass

$$x_{cm} = \frac{1}{m}\int x\,dm \quad , \quad y_{cm} = \frac{1}{m}\int y\,dm \quad , \quad z_{cm} = \frac{1}{m}\int z\,dm$$

Differential Mass Element

$dm = \lambda ds$	$dm = \sigma dA$	$dm = \rho dV$
(for an arc length, like a rod or circular arc)	(for a surface area, like a triangle or disc)	(for a 3D solid, like a solid cube or solid sphere)

Differential Arc Length

$ds = dx$	$ds = dy$	$ds = Rd\theta$
(along x)	(along y)	(circular arc)

Differential Area Element

$dA = dxdy$	$dA = rdrd\theta$	$dA = R^2 \sin\theta\, d\theta d\varphi$
(polygon)	(pie slice)	(spherical surface)

Differential Volume Element

$dV = dxdydz$	$dV = r_c dr_c d\theta dz$	$dV = r^2 \sin\theta\, drd\theta d\varphi$
(bounded by flat sides)	(cylinder or cone)	(spherical)

Symbol	Name	SI Units
dm	differential mass element	kg
m	total mass of the object	kg
x, y, z	Cartesian coordinates of dm	m, m, m
r, θ	2D polar coordinates of dm	m, rad
r_c, θ, z	cylindrical coordinates of dm	m, rad, m
r, θ, φ	spherical coordinates of dm	m, rad, rad
λ	linear mass density (for an arc)	kg/m
σ	mass density for a surface area	kg/m^2
ρ	mass density for a 3D solid	kg/m^3
ds	differential arc length	m
dA	differential area element	m^2
dV	differential volume element	m^3
x_{cm}	x-coordinate of the center of mass	m
y_{cm}	y-coordinate of the center of mass	m

Note: The symbols λ, σ, and ρ are the lowercase Greek letters lambda, sigma, and rho.

Cartesian Coordinates

2D Polar Coordinates

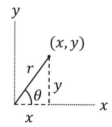

$$x = r\cos\theta \quad , \quad r = \sqrt{x^2 + y^2}$$
$$y = r\sin\theta \quad , \quad \theta = \tan^{-1}\left(\frac{y}{x}\right)$$

Cylindrical Coordinates

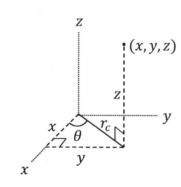

$$x = r_c\cos\theta \quad , \quad r_c = \sqrt{x^2 + y^2}$$
$$y = r_c\sin\theta \quad , \quad \theta = \tan^{-1}\left(\frac{y}{x}\right)$$

Spherical Coordinates

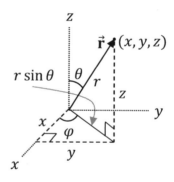

$$x = r\sin\theta\cos\varphi \quad , \quad r = \sqrt{x^2 + y^2 + z^2}$$
$$y = r\sin\theta\sin\varphi \quad , \quad \theta = \cos^{-1}\left(\frac{z}{r}\right)$$
$$z = r\cos\theta \quad , \quad \varphi = \tan^{-1}\left(\frac{y}{x}\right)$$

Example 169. A non-uniform rod has one end at the origin, the other end at $(0, 5.0 \text{ m})$, and non-uniform density $\lambda = \beta\sqrt{y}$, where β is a constant. Where is the center of mass of the rod?

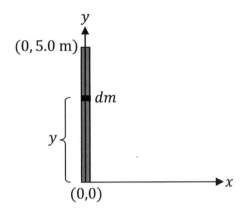

Solution. Begin with a labeled diagram. Draw a representative dm and label y for that dm, as we did in the diagram above. When we perform the integration, we effectively integrate over every dm that makes up the rod. Perform the center of mass integral.

$$y_{cm} = \frac{1}{m} \int y \, dm$$

Note that $x_{cm} = 0$ since the entire rod lies on the y-axis. For a rod, write $dm = \lambda ds$.

$$y_{cm} = \frac{1}{m} \int y \, (\lambda ds)$$

Since the rod lies on the y-axis, $ds = dy$. The limits of integration are from $y = 0$ to $y = 5$ m.

$$y_{cm} = \frac{1}{m} \int_{y=0}^{5} y\lambda \, dy$$

Don't pull λ out of the integral. Substitute $\lambda = \beta\sqrt{y}$.

$$y_{cm} = \frac{1}{m} \int_{y=0}^{5} y\left(\beta\sqrt{y}\right) dy$$

Apply the rule from algebra that $\sqrt{y} = y^{1/2}$.

$$y_{cm} = \frac{1}{m} \int_{y=0}^{5} y\left(\beta y^{1/2}\right) dy$$

Note that $yy^{1/2} = y^{3/2}$ according to the rule $y^m y^n = y^{m+n}$ since $1 + \frac{1}{2} = \frac{2}{2} + \frac{1}{2} = \frac{3}{2}$.

$$y_{cm} = \frac{1}{m} \int_{y=0}^{5} \beta y^{3/2} \, dy$$

Pull the constant β out of the integral.

$$y_{cm} = \frac{\beta}{m} \int_{y=0}^{5} y^{3/2} \, dy$$

Now we have an integral over y which we can perform using the method from Chapter 2.

$$y_{cm} = \frac{\beta}{m} \left[\frac{y^{3/2+1}}{3/2 + 1} \right]_{y=0}^{y=5}$$

To add fractions, use a **common denominator**: $\frac{3}{2} + 1 = \frac{3}{2} + \frac{2}{2} = \frac{5}{2}$. To divide by a fraction, multiply by its **reciprocal**.

$$y_{cm} = \frac{\beta}{m} \left[\frac{y^{5/2}}{5/2} \right]_{y=0}^{y=5} = \frac{\beta}{m} \left[\frac{2y^{5/2}}{5} \right]_{y=0}^{y=5} = \frac{\beta}{m} \left(\frac{2(5)^{5/2}}{5} - \frac{2(0)^{5/2}}{5} \right) = \frac{2\beta 5^{5/2}}{m5} = \frac{2\beta 5^{3/2}}{m}$$

Note that $\frac{5^{5/2}}{5} = \frac{5^{5/2}}{5^1} = 5^{5/2-1} = 5^{3/2}$ according to the rule $\frac{x^m}{x^n} = x^{m-n}$. In order to eliminate the mass (m) and the constant (β) from our answer, we will integrate over dm using the same substitutions as before, beginning with $dm = \lambda ds$.

$$m = \int dm = \int \lambda \, ds = \int_{y=0}^{5} \beta \sqrt{y} \, dy = \beta \int_{y=0}^{5} \sqrt{y} \, dy = \beta \int_{y=0}^{5} y^{1/2} \, dy$$

$$m = \beta \left[\frac{y^{1/2+1}}{1/2 + 1} \right]_{y=0}^{y=5}$$

To add fractions, use a **common denominator**: $\frac{1}{2} + 1 = \frac{1}{2} + \frac{2}{2} = \frac{3}{2}$. To divide by a fraction, multiply by its **reciprocal**.

$$m = \beta \left[\frac{y^{3/2}}{3/2} \right]_{y=0}^{y=5} = \beta \left[\frac{2y^{3/2}}{3} \right]_{y=0}^{y=5} = \beta \left(\frac{2(5)^{3/2}}{3} - \frac{2(0)^{3/2}}{3} \right) = \frac{2\beta 5^{3/2}}{3}$$

Substitute this expression for m into the previous expression for y_{cm}.

$$y_{cm} = \frac{2\beta 5^{3/2}}{m} = 2\beta 5^{3/2} \div \left(\frac{2\beta 5^{3/2}}{3} \right)$$

To divide by a fraction, multiply by its **reciprocal**.

$$y_{cm} = 2\beta 5^{3/2} \times \frac{3}{2\beta 5^{3/2}} = \frac{2\beta 5^{3/2}(3)}{2\beta 5^{3/2}} = 3.0 \text{ m}$$

The final answer is $y_{cm} = 3.0$ m. (This is 3 meters, **not** m for mass.) The location of the center of mass can also be expressed as $(x_{cm}, y_{cm}) = (0, 3.0 \text{ m})$. This result, $y_{cm} = 3.0$ m, should make sense conceptually. If the rod were uniform, the center of mass would be at its geometric center, 2.5 m from the origin. However, the rod is non-uniform. Since the non-uniform density is proportional to y (according to $\lambda = \beta\sqrt{y}$), there is more mass at the top of the rod (where y has a larger value). Therefore, the center of mass should be above the geometric center. Indeed, the center of mass $(y_{cm} = 3.0 \text{ m})$ is above the geometric center, as expected.

Example 170. Find the location of the center of mass of the uniform solid triangle illustrated below.

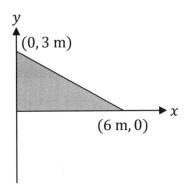

Solution. Begin with a labeled diagram. Draw a representative dm, as we did in the diagram below. When we perform the integration, we effectively integrate over every dm that makes up the solid triangle. Each dm has coordinates (x, y).

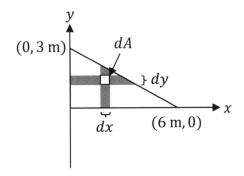

Perform the center of mass integrals.

$$x_{cm} = \frac{1}{m} \int x \, dm \quad , \quad y_{cm} = \frac{1}{m} \int y \, dm$$

For a solid triangle (as opposed to a wire bent into a triangle), we write $dm = \sigma dA$.

$$x_{cm} = \frac{1}{m} \int x\sigma \, dA \quad , \quad y_{cm} = \frac{1}{m} \int y\sigma \, dA$$

Since the triangle is uniform, σ is a constant and may come out of the integral.

$$x_{cm} = \frac{\sigma}{m} \int x \, dA \quad , \quad y_{cm} = \frac{\sigma}{m} \int y \, dA$$

For a triangle, we work with Cartesian coordinates and write the differential area element as $dA = dx dy$. This requires integrating over both x and y. If we let x vary from 0 to 6, observe that for a given value of x, y varies from 0 to the hypotenuse of the triangle, which is less than 3 (except when x reaches its upper limit). See the diagram above. To find the proper upper limit for y, we need the equation of the line that serves as the triangle's hypotenuse. That line has a slope equal to $\frac{y_2-y_1}{x_2-x_1} = \frac{0-3}{6-0} = -\frac{1}{2}$ and a y-intercept of 3. Since the general equation for a line is $y = mx + b$, the equation for this line is $y = -\frac{1}{2}x + 3$.

$$x_{cm} = \frac{\sigma}{m} \int\limits_{x=0}^{6} \int\limits_{y=0}^{-\frac{x}{2}+3} x \, dy \, dx \quad , \quad y_{cm} = \frac{\sigma}{m} \int\limits_{x=0}^{6} \int\limits_{y=0}^{-\frac{x}{2}+3} y \, dy \, dx$$

First integrate over y since the upper limit of y ($-\frac{x}{2} + 3$) is a variable. Factor x out of the y-integration (but not out of the x-integration).

$$x_{cm} = \frac{\sigma}{m} \int\limits_{x=0}^{6} x \left(\int\limits_{y=0}^{-\frac{x}{2}+3} dy \right) dx \quad , \quad y_{cm} = \frac{\sigma}{m} \int\limits_{x=0}^{6} \left(\int\limits_{y=0}^{-\frac{x}{2}+3} y \, dy \right) dx$$

$$x_{cm} = \frac{\sigma}{m} \int\limits_{x=0}^{6} x [y]_{y=0}^{y=-\frac{x}{2}+3} \, dx \quad , \quad y_{cm} = \frac{\sigma}{m} \int\limits_{x=0}^{6} \left[\frac{y^2}{2} \right]_{y=0}^{y=-\frac{x}{2}+3} dx$$

$$x_{cm} = \frac{\sigma}{m} \int\limits_{x=0}^{6} x \left(-\frac{x}{2} + 3 - 0 \right) dx \quad , \quad y_{cm} = \frac{\sigma}{m} \int\limits_{x=0}^{6} \left[\frac{\left(-\frac{x}{2} + 3 \right)^2}{2} - \frac{0^2}{2} \right] dx$$

Distribute in the first integral: $x \left(-\frac{x}{2} + 3 - 0 \right) = x \left(-\frac{x}{2} + 3 \right) = x \left(-\frac{x}{2} \right) + x(3) = -\frac{x^2}{2} + 3x$.

Factor out the $\frac{1}{2}$ in the second integral.

$$x_{cm} = \frac{\sigma}{m} \int\limits_{x=0}^{6} x \left(-\frac{x}{2} + 3 - 0 \right) dx \quad , \quad y_{cm} = \frac{\sigma}{2m} \int\limits_{x=0}^{6} \left[\left(-\frac{x}{2} + 3 \right)^2 \right] dx$$

Foil out the expression in the second integral: $\left(-\frac{x}{2} + 3 \right)^2 = \left(-\frac{x}{2} \right) \left(-\frac{x}{2} \right) + \left(-\frac{x}{2} \right) (3) +$

$(3) \left(-\frac{x}{2} \right) + (3)(3) = \frac{x^2}{4} - \frac{3x}{2} - \frac{3x}{2} + 9 = \frac{x^2}{4} - 3x + 9$.

$$x_{cm} = \frac{\sigma}{m} \int\limits_{x=0}^{6} \left(-\frac{x^2}{2} + 3x \right) dx \quad , \quad y_{cm} = \frac{\sigma}{2m} \int\limits_{x=0}^{6} \left(\frac{x^2}{4} - 3x + 9 \right) dx$$

Integrate using the technique from Chapter 2.

$$x_{cm} = \frac{\sigma}{m} \left[-\frac{x^3}{6} + \frac{3x^2}{2} \right]_{x=0}^{6} \quad , \quad y_{cm} = \frac{\sigma}{2m} \left[\frac{x^3}{12} - \frac{3x^2}{2} + 9x \right]_{x=0}^{6}$$

$$x_{cm} = \frac{\sigma}{m} \left[-\frac{(6)^3}{6} + \frac{(0)^3}{6} + \frac{3(6)^2}{2} - \frac{3(0)^2}{2} \right] = \frac{\sigma}{m} (-36 + 54) = \frac{18\sigma}{m}$$

$$y_{cm} = \frac{\sigma}{2m} \left[\frac{(6)^3}{12} - \frac{(0)^3}{12} - \frac{3(6)^2}{2} + \frac{3(0)^2}{2} + 9(6) - 9(0) \right] = \frac{\sigma}{2m} (18 - 54 + 54) = \frac{18\sigma}{2m} = \frac{9\sigma}{m}$$

In order to eliminate the mass (m) and the constant (σ) from our answer, we will integrate over dm using the same substitutions as before, beginning with $dm = \sigma dA$.

$$m = \int dm = \int \sigma \, dA = \sigma \int dA = \sigma \int\limits_{x=0}^{6} \int\limits_{y=0}^{-\frac{x}{2}+3} dy \, dx$$

$$m = \sigma \int_{x=0}^{6} \left(\int_{y=0}^{-\frac{x}{2}+3} dy \right) dx = \sigma \int_{x=0}^{6} [y]_{y=0}^{y=-\frac{x}{2}+3} dx = \sigma \int_{x=0}^{6} \left(-\frac{x}{2} + 3 - 0 \right) dx$$

$$m = \sigma \int_{x=0}^{6} \left(-\frac{x}{2} + 3 \right) dx = \sigma \left[-\frac{x^2}{4} + 3x \right]_{x=0}^{6} = \sigma \left[-\frac{(6)^2}{4} + \frac{(0)^2}{4} + 3(6) - 3(0) \right]$$

$$m = \sigma \left(-\frac{36}{4} + 18 \right) = \sigma(-9 + 18) = 9\sigma$$

Substitute this expression for m into the previous expressions for x_{cm} and y_{cm}.

$$x_{cm} = \frac{18\sigma}{m} = \frac{18\sigma}{9\sigma} = 2.0 \text{ m} \quad , \quad y_{cm} = \frac{9\sigma}{m} = \frac{9\sigma}{9\sigma} = 1.0 \text{ m}$$

The final answers are $x_{cm} = 2.0$ m and $y_{cm} = 1.0$ m. (These are meters, **not** mass.) The location of the center of mass can also be expressed as $(x_{cm}, y_{cm}) = (2.0 \text{ m}, 1.0 \text{ m})$. If you mark this location on the previous diagram, it should seem reasonable.

Example 171. A thin uniform ring with a 6.0-cm radius lies in the xy plane centered about the origin. A monkey cuts the thin ring in half along the x-axis and discards the negative y-portion, such that only a thin semicircular arc remains. Determine the location of its center of mass. **Note**: This is an arc length (such as one might make by bending a wire into one-half the circumference of a circle), and **not** a solid semicircle (such as would be made by cutting a coin in half).

Solution. Begin with a labeled diagram. Draw a representative dm, as we did in the diagram below. Each dm lies on the semicircular arc of radius $R = 6.0$ cm.

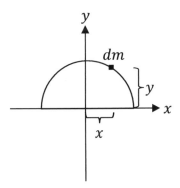

Perform the center of mass integral.

$$y_{cm} = \frac{1}{m} \int y \, dm$$

Note that $x_{cm} = 0$ due to symmetry about the y-axis. For a thin arc, write $dm = \lambda ds$.

$$y_{cm} = \frac{1}{m} \int y \lambda \, ds$$

Since the arc is uniform, λ is a constant and may come out of the integral.

$$y_{cm} = \frac{\lambda}{m} \int y \, ds$$

For a circular arc, work with 2D polar coordinates. The differential arc length is $ds = R d\theta$. The limits of integration are from $\theta = 0$ to $\theta = \pi$ **radians**. (The equation for arc length, $ds = R d\theta$, is only valid in radians; it doesn't work in degrees.)

$$y_{cm} = \frac{\lambda}{m} \int_{\theta=0}^{\pi} yR \, d\theta$$

Use the equation $y = R \sin \theta$ for 2D polar coordinates (from Chapter 26).

$$y_{cm} = \frac{\lambda}{m} \int_{\theta=0}^{\pi} R(R \sin \theta) \, d\theta = \frac{\lambda}{m} \int_{\theta=0}^{\pi} R^2 \sin \theta \, d\theta$$

You may pull R out of the integral since it is a very thin ring (not a solid semicircle). That is, the radius R is a constant (unlike the 2D polar coordinate r, which we'll encounter in the last example of this chapter).

$$y_{cm} = \frac{\lambda R^2}{m} \int_{\theta=0}^{\pi} \sin \theta \, d\theta$$

The anti-derivative of the sine function is the negative of the cosine function.

$$y_{cm} = \frac{\lambda R^2}{m} [-\cos \theta]_{\theta=0}^{\theta=\pi} = \frac{\lambda R^2}{m} [-\cos \pi - (-\cos 0)] = \frac{\lambda R^2}{m} [-(-1) - (-1)] = \frac{\lambda R^2}{m} (1+1)$$

$$y_{cm} = \frac{2\lambda R^2}{m}$$

In order to eliminate the mass (m) and the constant (λ) from our answer, we will integrate over dm using the same substitutions as before, beginning with $dm = \lambda ds$.

$$m = \int dm = \int \lambda \, ds = \lambda \int ds = \lambda \int_{\theta=0}^{\pi} R \, d\theta = \lambda R \int_{\theta=0}^{\pi} d\theta = \lambda R [\theta]_{\theta=0}^{\theta=\pi} = \lambda R (\pi - 0) = \pi \lambda R$$

Substitute this expression for m into the previous expression for y_{cm}.

$$y_{cm} = \frac{2\lambda R^2}{m} = \frac{2\lambda R^2}{\pi \lambda R} = \frac{2R}{\pi} = \frac{2(6)}{\pi} = \frac{12}{\pi} \text{ cm} = 3.8 \text{ cm}$$

The final answer is $y_{cm} = \frac{12}{\pi}$ cm $= 3.8$ cm. The location of the center of mass can also be expressed as $(x_{cm}, y_{cm}) = (0, \frac{12}{\pi}$ cm) or $(0, 3.8$ cm). If you think about the shape of the semicircular arc, it should make sense that the answer ($y_{cm} = 3.8$ cm) is higher than half the radius ($\frac{R}{2} = \frac{6}{2} = 3.0$ cm), since there is more mass above half the height than there is below half the height. (You may find it easier to visualize this if you mark the two points where y equals half the radius on the diagram.)

Example 172. Monkey Isle is the region bounded by the curve $y = x^2$ and the line $y = 4$. Monkey Isle has uniform density $\sigma = 2$. SI units have been suppressed. A monkey pirate buried a treasure in the center of mass of Monkey Isle. Find the (x, y) coordinates of the buried treasure.

Solution. Begin with a labeled diagram. Draw a representative dm, as we did in the diagram below. When we perform the integration, we effectively integrate over every dm that makes up the region. Each dm has coordinates (x, y).

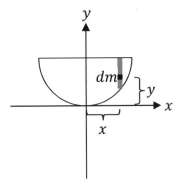

Perform the center of mass integral. Note that $x_{cm} = 0$ due to symmetry about the y-axis.

$$y_{cm} = \frac{1}{m} \int y \, dm$$

For a solid surface (the region between the parabola and the line $y = 4$ is a solid surface, as opposed to shaping a wire in the form of a parabola), we write $dm = \sigma dA$. Another way to realize this is that this problem gave us σ (instead of λ or ρ).

$$y_{cm} = \frac{1}{m} \int y\sigma \, dA$$

Since the region has uniform density, σ is a constant and may come out of the integral.

$$y_{cm} = \frac{\sigma}{m} \int y \, dA$$

For a parabola, we work with Cartesian coordinates and write the differential area element as $dA = dxdy$. This requires integrating over both x and y. If we let x vary from $x_{min} = -2$ to $x_{max} = 2$, observe that for a given value of x, y varies from $y_{min} = x^2$ to $y_{max} = 4$ (the vertical gray band shown in the diagram above). Note that letting x vary from -2 to 2 and letting y vary from 0 to 4 would be a mistake because that would make the region a rectangle. The lower limit of y will be x^2 (not 0) because the lower part of the region has the shape of a parabola.

$$y_{cm} = \frac{\sigma}{m} \int\limits_{x=-2}^{2} \int\limits_{y=x^2}^{4} y \, dy \, dx$$

How do we know that x varies from -2 to 2? The two points where the parabola $y = x^2$ intersects the horizontal line $y = 4$ can be found by setting $x^2 = 4$. Squareroot both sides to find $x = \sqrt{4}$. The two solutions are $x = -2$ and $x = 2$. Note that $(-2)^2 = (+2)^2 = 4$.

First integrate over y since the lower limit of y contains x.

$$y_{cm} = \frac{\sigma}{m} \int_{x=-2}^{2} \left(\int_{y=x^2}^{4} y \, dy \right) dx = \frac{\sigma}{m} \int_{x=-2}^{2} \left[\frac{y^2}{2} \right]_{y=x^2}^{y=4} dx = \frac{\sigma}{m} \int_{x=-2}^{2} \left[\frac{(4)^2}{2} - \frac{(x^2)^2}{2} \right] dx$$

$$= \frac{\sigma}{m} \int_{x=-2}^{2} \left[8 - \frac{x^4}{2} \right] dx = \frac{\sigma}{m} \left[8x - \frac{x^5}{10} \right]_{x=-2}^{2} = \frac{\sigma}{m} \left[8(2) - 8(-2) - \frac{(2)^5}{10} + \frac{(-2)^5}{10} \right]$$

Note that $(-2)^5 = -2^5 = -32$. To subtract fractions, find a **common denominator**.

$$y_{cm} = \frac{\sigma}{m} \left(16 + 16 - \frac{32}{10} + \frac{-32}{10} \right) = \frac{\sigma}{m} \left(32 - \frac{64}{10} \right) = \frac{\sigma}{m} \left(\frac{320}{10} - \frac{64}{10} \right) = \frac{\sigma}{m} \left(\frac{320 - 64}{10} \right) = \frac{256\sigma}{10m}$$

Divide both 256 and 10 by 2 to reduce the fraction: $\frac{256}{10} = \frac{256 \div 2}{10 \div 2} = \frac{128}{5}$.

$$y_{cm} = \frac{128\sigma}{5m}$$

The problem states that $\sigma = 2$. Plug this into the equation for the center of mass.

$$y_{cm} = \frac{128\sigma}{5m} = \frac{128(2)}{5m} = \frac{256}{5m}$$

In order to eliminate the mass (m) from our answer, we will integrate over dm using the same substitutions as before, beginning with $dm = \sigma dA$.

$$m = \int dm = \int \sigma \, dA = \sigma \int dA = \sigma \int_{x=-2}^{2} \int_{y=x^2}^{4} dy \, dx$$

$$m = \sigma \int_{x=-2}^{2} \left(\int_{y=x^2}^{4} dy \right) dx = \sigma \int_{x=-2}^{2} [y]_{y=x^2}^{y=4} dx = \sigma \int_{x=-2}^{2} (4 - x^2) \, dx$$

$$m = \sigma \left[4x - \frac{x^3}{3} \right]_{x=-2}^{2} = \sigma \left[4(2) - 4(-2) - \frac{(2)^3}{3} + \frac{(-2)^3}{3} \right]$$

$$m = \sigma \left(8 + 8 - \frac{8}{3} - \frac{8}{3} \right) = \sigma \left(16 - \frac{16}{3} \right) = \sigma \left(\frac{48}{3} - \frac{16}{3} \right) = \frac{32\sigma}{3}$$

Recall that $\sigma = 2$.

$$m = \frac{32\sigma}{3} = \frac{32(2)}{3} = \frac{64}{3} \text{ kg}$$

Substitute this expression for m into the previous expression y_{cm}.

$$y_{cm} = \frac{256}{5m} = \frac{256}{5} \div m = \frac{256}{5} \div \frac{64}{3} = \frac{256}{5} \times \frac{3}{64} = \frac{(256)(3)}{(5)(64)} = \frac{(4)(3)}{5} = \frac{12}{5} \text{ m}$$

The final answer is $y_{cm} = \frac{12}{5}$ m $= 2.4$ m. (These are meters, **not** mass.) The location of the center of mass can also be expressed as $(x_{cm}, y_{cm}) = (0, \frac{12}{5} \text{ m}) = (0, 2.4 \text{ m})$.

Example 173. The shaded semicircular ring illustrated below has an inner radius of 7.0 cm, an outer radius of 14.0 cm, and a non-uniform mass density $\sigma(r) = \beta r$, where β is a constant. Find the location of its center of mass.

Solution. Begin with a labeled diagram. Draw a representative dm, as we did in the diagram below. When we perform the integration, we effectively integrate over every dm that makes up the region. Each dm has coordinates (r, θ).

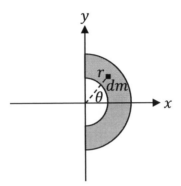

Perform the center of mass integral.

$$x_{cm} = \frac{1}{m} \int x \, dm$$

Note that $y_{cm} = 0$ due to symmetry about the x-axis. For a thick circular ring, we write $dm = \sigma dA$ (as opposed to a very thin ring, for which we would use λds instead).

$$x_{cm} = \frac{1}{m} \int x\sigma \, dA$$

The density is **non-uniform**, so it would be a mistake to pull σ out of the integral. Instead, substitute the equation given in the problem, $\sigma = \beta r$, in place of σ.

$$x_{cm} = \frac{1}{m} \int x(\beta r) \, dA$$

Pull the constant β out of the integral (but don't pull x or r out of the integral, as they aren't constants: their values are different for different dm's).

$$x_{cm} = \frac{\beta}{m} \int xr \, dA$$

For a circular ring, the math is simpler if we work with 2D polar coordinates. The differential area element is $dA = rdrd\theta$. The limits of integration are from $r = 7$ cm to $r = 14$ cm and $\theta = -\frac{\pi}{2}$ to $\theta = \frac{\pi}{2}$ **radians**. (Don't use degrees.) Note that 270° is the same as −90°, and note that 90° equates to $\frac{\pi}{2}$ radians (since 360° is equal to 2π radians).

$$x_{cm} = \frac{\beta}{m} \int_{r=7}^{14} \int_{\theta=-\frac{\pi}{2}}^{\frac{\pi}{2}} xr \, (rdr \, d\theta) = \frac{\beta}{m} \int_{r=7}^{14} \int_{\theta=-\frac{\pi}{2}}^{\frac{\pi}{2}} xr^2 \, dr \, d\theta$$

Use the equation $x = r \cos \theta$ for 2D polar coordinates (from Chapter 26).

$$x_{cm} = \frac{\beta}{m} \int\limits_{r=7}^{14} \int\limits_{\theta=-\frac{\pi}{2}}^{\frac{\pi}{2}} (r\cos\theta)r^2\, dr\, d\theta = \frac{\beta}{m} \int\limits_{r=7}^{14} \int\limits_{\theta=-\frac{\pi}{2}}^{\frac{\pi}{2}} r^3\cos\theta\, dr\, d\theta$$

You may do these integrals in any order since the limits are constant. We will integrate over θ first. Factor r out of the θ-integration (but not the r-integration). Recall from calculus that the anti-derivative of the cosine function is the sine function.

$$x_{cm} = \frac{\beta}{m} \int\limits_{r=7}^{14} r^3 \left(\int\limits_{\theta=-\frac{\pi}{2}}^{\frac{\pi}{2}} \cos\theta\, d\theta \right) dr = \frac{\beta}{m} \int\limits_{r=7}^{14} r^3 [\sin\theta]_{\theta=-\frac{\pi}{2}}^{\theta=\frac{\pi}{2}} dr$$

$$x_{cm} = \frac{\beta}{m} \int\limits_{r=7}^{14} r^3 \left[\sin\left(\frac{\pi}{2}\right) - \sin\left(-\frac{\pi}{2}\right)\right]_{\theta=-\frac{\pi}{2}}^{\theta=\frac{\pi}{2}} dr = \frac{\beta}{m} \int\limits_{r=7}^{14} r^3[1-(-1)]\, dr = \frac{\beta}{m} \int\limits_{r=7}^{14} 2r^3\, dr$$

Note that $\sin\left(\frac{\pi}{2}\right) = \sin(90°) = 1$ and $\sin\left(-\frac{\pi}{2}\right) = \sin(-90°) = \sin(270°) = -1$.

$$x_{cm} = \frac{\beta}{m} \int\limits_{r=7}^{14} 2r^3\, dr = \frac{\beta}{m}\left[\frac{2r^4}{4}\right]_{r=7}^{r=14} = \frac{\beta}{m}\left[\frac{r^4}{2}\right]_{r=7}^{r=14} = \frac{\beta}{m}\left[\frac{(14)^4}{2} - \frac{(7)^4}{2}\right]$$

$$x_{cm} = \frac{\beta}{m}\left(\frac{38,416}{2} - \frac{2401}{2}\right) = \frac{\beta}{m}\left(\frac{36,015}{2}\right) = \frac{36,015\beta}{2m}$$

In order to eliminate the mass (m) from our answer, we will integrate over dm using the same substitutions as before, beginning with $dm = \sigma dA$.

$$m = \int dm = \int \sigma\, dA = \int \beta r\, dA = \beta \int r\, dA = \beta \int\limits_{r=7}^{14} \int\limits_{\theta=-\frac{\pi}{2}}^{\frac{\pi}{2}} r\,(r dr\, d\theta) = \beta \int\limits_{r=7}^{14} \int\limits_{\theta=-\frac{\pi}{2}}^{\frac{\pi}{2}} r^2\, dr\, d\theta$$

$$m = \beta \int\limits_{r=7}^{14} r^2 \left(\int\limits_{\theta=-\frac{\pi}{2}}^{\frac{\pi}{2}} d\theta \right) dr = \beta \int\limits_{r=7}^{14} r^2[\theta]_{\theta=-\frac{\pi}{2}}^{\theta=\frac{\pi}{2}} dr = \beta \int\limits_{r=7}^{14} r^2 \left[\frac{\pi}{2} - \left(-\frac{\pi}{2}\right)\right] dr$$

$$m = \beta \int\limits_{r=7}^{14} r^2 \left(\frac{\pi}{2} + \frac{\pi}{2}\right) dr = \beta \int\limits_{r=7}^{14} r^2(\pi)dr = \beta \int\limits_{r=7}^{14} \pi r^2\, dr = \beta \left[\frac{\pi r^3}{3}\right]_{r=7}^{r=14}$$

$$m = \beta \left[\frac{\pi(14)^3}{3} - \frac{\pi(7)^3}{3}\right] = \beta \left[\frac{2744\pi}{3} - \frac{343\pi}{3}\right] = \frac{2401\pi\beta}{3}$$

Substitute this expression for m into the previous expression y_{cm}.

$$x_{cm} = \frac{36,015\beta}{2} \div m = \frac{36,015\beta}{2} \div \frac{2401\pi\beta}{3} = \frac{36,015\beta}{2} \times \frac{3}{2401\pi\beta} = \frac{45}{2\pi} \text{ cm} = 7.2 \text{ cm}$$

The final answer is $x_{cm} = \frac{45}{2\pi}$ cm $= 7.2$ cm. (These are meters, **not** mass.) The location of the center of mass can also be expressed as $(x_{cm}, y_{cm}) = \left(\frac{45}{2\pi} \text{ cm}, 0\right) = (7.2 \text{ cm}, 0)$.

28 UNIFORM ANGULAR ACCELERATION

Uniform Angular Acceleration

$$\Delta\theta = \omega_0 t + \frac{1}{2}\alpha t^2 \quad , \quad \omega = \omega_0 + \alpha t \quad , \quad \omega^2 = \omega_0^2 + 2\alpha\Delta\theta$$

Uniform Tangential Acceleration

$$\Delta s = v_{T0}t + \frac{1}{2}a_T t^2 \quad , \quad v_T = v_{T0} + a_T t \quad , \quad v_T^2 = v_{T0}^2 + 2a_T\Delta s$$

Angular/Tangential Relations

$$\Delta s = R\Delta\theta \quad , \quad v_T = R\omega \quad , \quad a_T = R\alpha$$

Calculus Relations between Angular Variables

$$\omega = \frac{d\theta}{dt} \quad , \quad \alpha = \frac{d\omega}{dt}$$

Centripetal Acceleration

$$a_c = \frac{v^2}{R}$$

Magnitude of the Total Acceleration

$$a = \sqrt{a_T^2 + a_c^2}$$

Symbol	Name	SI Units
$\Delta\theta$	angular displacement	rad
ω_0	initial angular velocity	rad/s
ω	final angular velocity	rad/s
α	angular acceleration	rad/s^2
t	time	s
Δs	arc length	m
v_{T0}	initial tangential velocity	m/s
v_T	final tangential velocity	m/s
a_T	tangential acceleration	m/s^2
a_c	centripetal acceleration	m/s^2
a	acceleration	m/s^2

Note: The symbols θ, ω, and α are the lowercase Greek letters theta, omega, and alpha.

Name	Definition
angular acceleration	the instantaneous rate at which angular velocity is changing
tangential acceleration	the instantaneous rate at which speed is changing
centripetal acceleration	the instantaneous rate at which the direction is changing
acceleration	the instantaneous rate at which velocity is changing

Example 174. A monkey drives a bananamobile in a circle with an initial angular speed of 8.0 rev/s and a uniform angular acceleration of 16.0 rev/s², completing 6.0 revolutions.
(A) How much time does this take?

Solution. Begin by making a list of the known quantities.

- The initial angular speed is $\omega_0 = 8.0$ rev/s.
- The angular acceleration is $\alpha = 16.0$ rev/s².
- The angular displacement is $\Delta\theta = 6.0$ rev.

It's okay to leave everything in revolutions if you're working exclusively with angular variables. (If instead you need to use a tangential equation, such as $v_T = R\omega$, then you must convert to radians first.) Since we know $\Delta\theta$, ω_0, and α, and since we're solving for t, we should use an equation that only has these four symbols.

$$\Delta\theta = \omega_0 t + \frac{1}{2}\alpha t^2$$

Plug the knowns into this equation. To avoid clutter, suppress the units until the end.

$$6 = 8t + \frac{1}{2}(16)t^2$$

Simplify this equation.

$$6 = 8t + 8t^2$$

Recognize that this is a **quadratic equation** because it includes a quadratic term ($8t^2$), a linear term ($8t$), and a constant term (6). Use algebra to bring the constant (6) to the right-hand side, so that all three terms are on the same side of the equation. (This term will change sign when we subtract 6 from both sides.) Also, order the terms such that the equation is in **standard form**, with the quadratic term first, the linear term second, and the constant term last.

$$0 = 8t^2 + 8t - 6$$

Compare this equation to the general form $at^2 + bt + c = 0$ to identify the constants.

$$a = 8 \quad , \quad b = 8 \quad , \quad c = -6$$

Plug these constants into the **quadratic formula**.

$$t = \frac{-b \pm \sqrt{b^2 - 4ac}}{2a} = \frac{-8 \pm \sqrt{(8)^2 - 4(8)(-6)}}{2(8)}$$

Note that two minus signs make a plus sign: $-4(8)(-6) = +192$.

$$t = \frac{-8 \pm \sqrt{64 + 192}}{16} = \frac{-8 \pm \sqrt{256}}{16} = \frac{-8 \pm 16}{16}$$

We must consider both solutions. Work out the two cases separately.

$$t = \frac{-8 + 16}{16} \quad \text{or} \quad t = \frac{-8 - 16}{16}$$

$$t = \frac{8}{16} \quad \text{or} \quad t = \frac{-24}{16}$$

$$t = 0.50 \text{ s} \quad \text{or} \quad t = -1.5 \text{ s}$$

Since time can't be negative, the correct answer is $t = 0.50$ s, which is the same as $t = \frac{1}{2}$ s.

(B) What is the final angular speed?

Solution. Solve for ω.

$$\omega = \omega_0 + \alpha t = 8 + (16)(0.5) = 8 + 8 = 16 \text{ rev/s}$$

The final angular speed is 16 rev/s.

Example 175. A monkey drives a bananamobile in a circle with an initial angular speed of $\frac{1}{5}$ rev/s and a uniform angular acceleration of $\frac{1}{20}$ rev/s² for one minute.

(A) How many revolutions does the bananamobile complete?

Solution. Begin by making a list of the known quantities.

- The initial angular speed is $\omega_0 = \frac{1}{5}$ rev/s.
- The angular acceleration is $\alpha = \frac{1}{20}$ rev/s².
- The time is $t = 60$ s (after converting the minute to seconds).

Since we know ω_0, α, and t, and since we're solving for $\Delta\theta$, we should use an equation that only has these four symbols. Since we only need to use an angular equation (and don't need to work with tangential quantities), we may leave everything in revolutions.

$$\Delta\theta = \omega_0 t + \frac{1}{2}\alpha t^2 = \left(\frac{1}{5}\right)(60) + \frac{1}{2}\left(\frac{1}{20}\right)(60)^2 = 12 + \frac{3600}{40} = 12 + 90 = 102 \text{ rev}$$

The number of revolutions completed is $\Delta\theta = 102$ rev.

(B) If the radius of the circle is 15 m, what is the final speed of the bananamobile?

Solution. First find the final angular speed.

$$\omega = \omega_0 + \alpha t = \frac{1}{5} + \left(\frac{1}{20}\right)(60) = \frac{1}{5} + 3 = \frac{1}{5} + \frac{15}{5} = \frac{16}{5} \text{ rev/s} = 3.2 \text{ rev/s}$$

Now we need to use a tangential quantity (v_T). Therefore, we can no longer work with revolutions. We must convert the angular speed to **radians**. Note that 1 rev $= 2\pi$ rad.

$$\omega = \frac{16}{5} \text{ rev/s} = \frac{16}{5} \text{ rev/s} \times \frac{2\pi \text{ rad}}{1 \text{ rev}} = \frac{32\pi}{5} \text{ rad/s} = 20.1 \text{ rad/s}$$

Use the equation that relates tangential velocity to angular speed.

$$v_T = R\omega = (15)\left(\frac{32\pi}{5}\right) = 96\pi \text{ m/s} = 302 \text{ m/s}$$

Since speed is the magnitude of velocity, the final speed is $v = 96\pi$ m/s $= 302$ m/s.

29 THE VECTOR PRODUCT

Equations	
$\begin{vmatrix} a & b \\ c & d \end{vmatrix} = ad - bc$	$\|\vec{\mathbf{A}} \times \vec{\mathbf{B}}\| = AB \sin\theta$

$$\begin{vmatrix} a & b & c \\ d & e & f \\ g & h & i \end{vmatrix} = a\begin{vmatrix} e & f \\ h & i \end{vmatrix} - b\begin{vmatrix} d & f \\ g & i \end{vmatrix} + c\begin{vmatrix} d & e \\ g & h \end{vmatrix}$$

$$= aei - afh - bdi + bfg + cdh - ceg$$

$$\begin{vmatrix} a & b & c \\ d & e & f \\ g & h & i \end{vmatrix}\begin{matrix} a & b \\ d & e \\ g & h \end{matrix} = aei + bfg + cdh - ceg - afh - bdi$$

$$\vec{\mathbf{A}} \times \vec{\mathbf{B}} = \begin{vmatrix} \hat{\mathbf{x}} & \hat{\mathbf{y}} & \hat{\mathbf{z}} \\ A_x & A_y & A_z \\ B_x & B_y & B_z \end{vmatrix} = \hat{\mathbf{x}}\begin{vmatrix} A_y & A_z \\ B_y & B_z \end{vmatrix} - \hat{\mathbf{y}}\begin{vmatrix} A_x & A_z \\ B_x & B_z \end{vmatrix} + \hat{\mathbf{z}}\begin{vmatrix} A_x & A_y \\ B_x & B_y \end{vmatrix}$$

$$\vec{\mathbf{A}} \times \vec{\mathbf{B}} = (A_yB_z - A_zB_y)\hat{\mathbf{x}} - (A_xB_z - A_zB_x)\hat{\mathbf{y}} + (A_xB_y - A_yB_x)\hat{\mathbf{z}}$$

$$\vec{\mathbf{A}} \times \vec{\mathbf{B}} = A_yB_z\hat{\mathbf{x}} - A_zB_y\hat{\mathbf{x}} + A_zB_x\hat{\mathbf{y}} - A_xB_z\hat{\mathbf{y}} + A_xB_y\hat{\mathbf{z}} - A_yB_x\hat{\mathbf{z}}$$

$\vec{\mathbf{A}} = A_x\hat{\mathbf{x}} + A_y\hat{\mathbf{y}} + A_z\hat{\mathbf{z}}$	$\vec{\mathbf{B}} = B_x\hat{\mathbf{x}} + B_y\hat{\mathbf{y}} + B_z\hat{\mathbf{z}}$
$A = \sqrt{A_x^2 + A_y^2 + A_z^2}$	$B = \sqrt{B_x^2 + B_y^2 + B_z^2}$

Vector Product between Unit Vectors		
$\hat{\mathbf{x}} \times \hat{\mathbf{y}} = \hat{\mathbf{z}}$	$\hat{\mathbf{y}} \times \hat{\mathbf{z}} = \hat{\mathbf{x}}$	$\hat{\mathbf{z}} \times \hat{\mathbf{x}} = \hat{\mathbf{y}}$
$\hat{\mathbf{y}} \times \hat{\mathbf{x}} = -\hat{\mathbf{z}}$	$\hat{\mathbf{z}} \times \hat{\mathbf{y}} = -\hat{\mathbf{x}}$	$\hat{\mathbf{x}} \times \hat{\mathbf{z}} = -\hat{\mathbf{y}}$
$\hat{\mathbf{x}} \times \hat{\mathbf{x}} = 0$	$\hat{\mathbf{y}} \times \hat{\mathbf{y}} = 0$	$\hat{\mathbf{z}} \times \hat{\mathbf{z}} = 0$

Example 176. Find the vector product, $\vec{A} \times \vec{B}$, between the following vectors.

$$\vec{A} = 5\,\hat{x} + 2\,\hat{y} + 4\,\hat{z}$$
$$\vec{B} = 3\,\hat{x} + 4\,\hat{y} + 6\,\hat{z}$$

Solution. First identify the components of the given vectors.

- Compare $\vec{A} = 5\,\hat{x} + 2\,\hat{y} + 4\,\hat{z}$ to $\vec{A} = A_x\hat{x} + A_y\hat{y} + A_z\hat{z}$. The components of vector $\vec{A}$ are $A_x = 5$, $A_y = 2$, and $A_z = 4$.

- Compare $\vec{B} = 3\,\hat{x} + 4\,\hat{y} + 6\,\hat{z}$ to $\vec{B} = B_x\hat{x} + B_y\hat{y} + B_z\hat{z}$. The components of vector $\vec{B}$ are $B_x = 3$, $B_y = 4$, and $B_z = 6$.

Use the determinant form of the vector product.

$$\vec{A} \times \vec{B} = \begin{vmatrix} \hat{x} & \hat{y} & \hat{z} \\ A_x & A_y & A_z \\ B_x & B_y & B_z \end{vmatrix} = \hat{x}\begin{vmatrix} A_y & A_z \\ B_y & B_z \end{vmatrix} - \hat{y}\begin{vmatrix} A_x & A_z \\ B_x & B_z \end{vmatrix} + \hat{z}\begin{vmatrix} A_x & A_y \\ B_x & B_y \end{vmatrix}$$

$$\vec{A} \times \vec{B} = \begin{vmatrix} \hat{x} & \hat{y} & \hat{z} \\ 5 & 2 & 4 \\ 3 & 4 & 6 \end{vmatrix} = \hat{x}\begin{vmatrix} 2 & 4 \\ 4 & 6 \end{vmatrix} - \hat{y}\begin{vmatrix} 5 & 4 \\ 3 & 6 \end{vmatrix} + \hat{z}\begin{vmatrix} 5 & 2 \\ 3 & 4 \end{vmatrix}$$

$$\vec{A} \times \vec{B} = [(2)(6) - (4)(4)]\hat{x} - [(5)(6) - (4)(3)]\hat{y} + [(5)(4) - (2)(3)]\hat{z}$$

$$\vec{A} \times \vec{B} = (12 - 16)\hat{x} - (30 - 12)\hat{y} + (20 - 6)\hat{z} = -4\,\hat{x} - 18\,\hat{y} + 14\,\hat{z}$$

The vector product is $\vec{A} \times \vec{B} = -4\,\hat{x} - 18\,\hat{y} + 14\,\hat{z}$.

Example 177. Find the vector product, $\vec{\mathbf{A}} \times \vec{\mathbf{B}}$, between the following vectors.

$$\vec{\mathbf{A}} = 3\,\hat{\mathbf{x}} - 2\,\hat{\mathbf{z}}$$
$$\vec{\mathbf{B}} = 8\,\hat{\mathbf{x}} - \hat{\mathbf{y}} - 4\,\hat{\mathbf{z}}$$

Solution. First identify the components of the given vectors.

- Compare $\vec{\mathbf{A}} = 3\,\hat{\mathbf{x}} - 2\,\hat{\mathbf{z}}$ to $\vec{\mathbf{A}} = A_x\hat{\mathbf{x}} + A_y\hat{\mathbf{y}} + A_z\hat{\mathbf{z}}$. The components of vector $\vec{\mathbf{A}}$ are $A_x = 3$, $A_y = 0$, and $A_z = -2$. Observe that $\vec{\mathbf{A}}$ doesn't have a $\hat{\mathbf{y}}$: That's why $A_y = 0$.
- Compare $\vec{\mathbf{B}} = 8\,\hat{\mathbf{x}} - \hat{\mathbf{y}} - 4\,\hat{\mathbf{z}}$ to $\vec{\mathbf{B}} = B_x\hat{\mathbf{x}} + B_y\hat{\mathbf{y}} + B_z\hat{\mathbf{z}}$. The components of vector $\vec{\mathbf{B}}$ are $B_x = 8$, $B_y = -1$, and $B_z = -4$.

Use the determinant form of the vector product.

$$\vec{\mathbf{A}} \times \vec{\mathbf{B}} = \begin{vmatrix} \hat{\mathbf{x}} & \hat{\mathbf{y}} & \hat{\mathbf{z}} \\ A_x & A_y & A_z \\ B_x & B_y & B_z \end{vmatrix} = \hat{\mathbf{x}}\begin{vmatrix} A_y & A_z \\ B_y & B_z \end{vmatrix} - \hat{\mathbf{y}}\begin{vmatrix} A_x & A_z \\ B_x & B_z \end{vmatrix} + \hat{\mathbf{z}}\begin{vmatrix} A_x & A_y \\ B_x & B_y \end{vmatrix}$$

$$\vec{\mathbf{A}} \times \vec{\mathbf{B}} = \begin{vmatrix} \hat{\mathbf{x}} & \hat{\mathbf{y}} & \hat{\mathbf{z}} \\ 3 & 0 & -2 \\ 8 & -1 & -4 \end{vmatrix} = \hat{\mathbf{x}}\begin{vmatrix} 0 & -2 \\ -1 & -4 \end{vmatrix} - \hat{\mathbf{y}}\begin{vmatrix} 3 & -2 \\ 8 & -4 \end{vmatrix} + \hat{\mathbf{z}}\begin{vmatrix} 3 & 0 \\ 8 & -1 \end{vmatrix}$$

$$\vec{\mathbf{A}} \times \vec{\mathbf{B}} = [(0)(-4) - (-2)(-1)]\hat{\mathbf{x}} - [(3)(-4) - (-2)(8)]\hat{\mathbf{y}} + [(3)(-1) - (0)(8)]\hat{\mathbf{z}}$$

$$\vec{\mathbf{A}} \times \vec{\mathbf{B}} = (0 - 2)\hat{\mathbf{x}} - [-12 - (-16)]\hat{\mathbf{y}} + (-3 - 0)\hat{\mathbf{z}} = -2\,\hat{\mathbf{x}} - (-12 + 16)\hat{\mathbf{y}} - 3\,\hat{\mathbf{z}}$$

$$\vec{\mathbf{A}} \times \vec{\mathbf{B}} = -2\,\hat{\mathbf{x}} - 4\,\hat{\mathbf{y}} - 3\,\hat{\mathbf{z}}$$

The vector product is $\vec{\mathbf{A}} \times \vec{\mathbf{B}} = -2\,\hat{\mathbf{x}} - 4\,\hat{\mathbf{y}} - 3\,\hat{\mathbf{z}}$.

Example 178. Given $\vec{A} = -\hat{y} + \hat{z}$ and $\vec{B} = \hat{x} - \hat{z}$, find $\vec{C}$ where $\vec{C} = \vec{A} \times \vec{B}$.

Solution. First identify the components of the given vectors.

- Compare $\vec{A} = -\hat{y} + \hat{z}$ to $\vec{A} = A_x\hat{x} + A_y\hat{y} + A_z\hat{z}$. The components of vector $\vec{A}$ are $A_x = 0$, $A_y = -1$, and $A_z = 1$. Observe that $\vec{A}$ doesn't have an $\hat{x}$: That's why $A_x = 0$.

- Compare $\vec{B} = \hat{x} - \hat{z}$ to $\vec{B} = B_x\hat{x} + B_y\hat{y} + B_z\hat{z}$. The components of vector $\vec{B}$ are $B_x = 1$, $B_y = 0$, and $B_z = -1$. Observe that $\vec{B}$ doesn't have a $\hat{y}$: That's why $B_y = 0$.

Use the determinant form of the vector product.

$$\vec{A} \times \vec{B} = \begin{vmatrix} \hat{x} & \hat{y} & \hat{z} \\ A_x & A_y & A_z \\ B_x & B_y & B_z \end{vmatrix} = \hat{x}\begin{vmatrix} A_y & A_z \\ B_y & B_z \end{vmatrix} - \hat{y}\begin{vmatrix} A_x & A_z \\ B_x & B_z \end{vmatrix} + \hat{z}\begin{vmatrix} A_x & A_y \\ B_x & B_y \end{vmatrix}$$

$$\vec{A} \times \vec{B} = \begin{vmatrix} \hat{x} & \hat{y} & \hat{z} \\ 0 & -1 & 1 \\ 1 & 0 & -1 \end{vmatrix} = \hat{x}\begin{vmatrix} -1 & 1 \\ 0 & -1 \end{vmatrix} - \hat{y}\begin{vmatrix} 0 & 1 \\ 1 & -1 \end{vmatrix} + \hat{z}\begin{vmatrix} 0 & -1 \\ 1 & 0 \end{vmatrix}$$

$$\vec{A} \times \vec{B} = [(-1)(-1) - (1)(0)]\hat{x} - [(0)(-1) - (1)(1)]\hat{y} + [(0)(0) - (-1)(1)]\hat{z}$$

$$\vec{A} \times \vec{B} = (1 - 0)\hat{x} - (0 - 1)\hat{y} + [0 - (-1)]\hat{z} = \hat{x} - (-1)\hat{y} + \hat{z}$$

$$\vec{A} \times \vec{B} = \hat{x} + \hat{y} + \hat{z}$$

The vector product is $\vec{C} = \vec{A} \times \vec{B} = \hat{x} + \hat{y} + \hat{z}$.

Example 179. $\vec{A}$ has a magnitude of 3 and $\vec{B}$ has a magnitude of 4. The angle between $\vec{A}$ and $\vec{B}$ is 150°. Find the magnitude of $\vec{C}$, where $\vec{C} = \vec{A} \times \vec{B}$.

Solution. Use the trigonometric form of the vector product.

$$C = \|\vec{A} \times \vec{B}\| = AB \sin\theta$$

$$C = (3)(4)\sin 150° = 12\left(\frac{1}{2}\right) = 6$$

The magnitude of $\vec{C}$ is $C = \|\vec{A} \times \vec{B}\| = 6$.

30 TORQUE

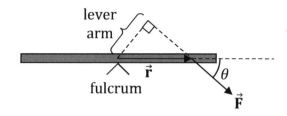

Equations	
$\vec{\boldsymbol{\tau}} = \vec{\mathbf{r}} \times \vec{\mathbf{F}}$	$\tau = \|\vec{\mathbf{r}} \times \vec{\mathbf{F}}\| = r\,F\sin\theta$

$$\vec{\boldsymbol{\tau}} = \vec{\mathbf{r}} \times \vec{\mathbf{F}} = \begin{vmatrix} \hat{\mathbf{x}} & \hat{\mathbf{y}} & \hat{\mathbf{z}} \\ x & y & z \\ F_x & F_y & F_z \end{vmatrix} = \hat{\mathbf{x}}\begin{vmatrix} y & z \\ F_y & F_z \end{vmatrix} - \hat{\mathbf{y}}\begin{vmatrix} x & z \\ F_x & F_z \end{vmatrix} + \hat{\mathbf{z}}\begin{vmatrix} x & y \\ F_x & F_y \end{vmatrix}$$
$$\vec{\boldsymbol{\tau}} = (yF_z - zF_y)\hat{\mathbf{x}} - (xF_z - zF_x)\hat{\mathbf{y}} + (xF_y - yF_x)\hat{\mathbf{z}}$$
$$\vec{\boldsymbol{\tau}} = yF_z\hat{\mathbf{x}} - zF_y\hat{\mathbf{x}} + zF_x\hat{\mathbf{y}} - xF_z\hat{\mathbf{y}} + xF_y\hat{\mathbf{z}} - yF_x\hat{\mathbf{z}}$$

$\vec{\mathbf{r}} = x\hat{\mathbf{x}} + y\hat{\mathbf{y}} + z\hat{\mathbf{z}}$	$\vec{\mathbf{F}} = F_x\hat{\mathbf{x}} + F_y\hat{\mathbf{y}} + F_z\hat{\mathbf{z}}$
lever arm $= r\sin\theta$	$\tau = \sqrt{\tau_x^2 + \tau_y^2 + \tau_z^2}$

Symbol	Quantity	Units
r	distance from the axis of rotation to the point where the force is applied	m
F	force	N
θ	the angle between $\vec{\mathbf{r}}$ and $\vec{\mathbf{F}}$	° or rad
τ	torque	Nm

Note: The symbol for torque (τ) is the lowercase Greek letter tau.

Example 180. As illustrated below, a 30-kg white box is at the very right end of a 60-kg plank. The plank is 20.0 m long and the fulcrum is 4.0-m from the left end.

(A) Find the torque exerted on the system due to the weight of the white box.

Solution. Draw a picture showing $\vec{r}_1$ and $\vec{F}_1$. The specified force is the weight ($m_1\vec{g}$) of the white box, while $\vec{r}_1$ extends from the fulcrum to the white box.

Identify the quantities involved in the torque equation.

- $\vec{r}_1$ extends from the fulcrum to the white box.
$$r_1 = L - 4 = 20 - 4 = 16.0 \text{ m}$$
- The force is the weight of the white box: $F_1 = m_1 g$. The mass of the white box is $m_1 = 30$ kg. The weight of the white box is $F_1 = m_1 g = (30)(9.81) = 294$ N.
- The angle is $\theta_1 = 90°$ since $\vec{r}_1$ (horizontal) and $\vec{F}_1$ (vertical) are perpendicular.

Plug this information into the equation for the magnitude of the torque.
$$\tau_1 = r_1 F_1 \sin\theta_1 = (16)(294)\sin 90° = 4704 \text{ Nm} = 4.70 \times 10^3 \text{ Nm}$$
The torque is $\tau_1 = 4704$ Nm $= 4.70 \times 10^3$ Nm. (If you round gravity to ≈ 10 m/s², then the torque is approximately $\tau_1 \approx 4800$ Nm.)

(B) Find the torque exerted on the system due to the weight of the plank.

Solution. Draw a picture showing $\vec{r}_2$ and $\vec{F}_2$. The specified force is the weight ($m_2\vec{g}$) of the plank, while $\vec{r}_2$ extends from the fulcrum to the center of the plank. (The center of the plank is the point where gravity acts on **average**).

Identify the quantities involved in the torque equation.

- $\vec{r}_2$ extends from the fulcrum to the center of the plank.
$$r_2 = \frac{L}{2} - 4 = \frac{20}{2} - 4 = 10 - 4 = 6.0 \text{ m}$$
- The force is the weight of the plank: $F_2 = m_2 g$. The mass of the plank is $m_2 = 60$ kg. The weight of the plank is $F_2 = m_2 g = (60)(9.81) = 589$ N.
- The angle is $\theta_2 = 90°$ since $\vec{r}_2$ (horizontal) and $\vec{F}_2$ (vertical) are perpendicular.

Plug this information into the equation for the magnitude of the torque.
$$\tau_2 = r_2 F_2 \sin\theta_2 = (6)(589)\sin 90° = 3534 \text{ Nm} = 3.53 \times 10^3 \text{ Nm}$$
The torque is $\tau_2 = 3534$ Nm $= 3.53 \times 10^3$ Nm. (If you round gravity to ≈ 10 m/s², then the torque is approximately $\tau_2 \approx 3600$ Nm.)

Example 181. As illustrated below, one end of a 6.0-m long, 7.0-kg rod is connected to a wall by a hinge, which allows the rod to rotate. (The rod won't stay in the position shown: It will rotate until it reaches the wall, where the rod will be vertical.) Find the instantaneous torque exerted on the rod due to the weight of the rod in the position shown below.

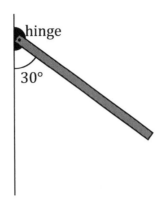

Solution. Draw a picture showing $\vec{r}$ and $\vec{F}$. The specified force is the weight ($m\vec{g}$) of the rod, while $\vec{r}$ extends from the hinge to the center of the rod. (The center of the rod is the point where gravity acts on **average**).

Identify the quantities involved in the torque equation.

- $\vec{r}$ extends from the hinge to the center of the rod.

$$r = \frac{L}{2} = \frac{6}{2} = 3.0 \text{ m}$$

- The force is the weight of the rod: $F = mg$. The mass of the rod is $m = 7.0$ kg. The weight of the rod is $F = mg = (7)(9.81) = 69$ N.
- The angle between $\vec{r}$ (along the rod) and $\vec{F}$ (straight down) is $\theta = 30°$.

Plug this information into the equation for the magnitude of the torque.

$$\tau = rF \sin \theta = (3)(69) \sin 30° = (207)\left(\frac{1}{2}\right) = 104 \text{ Nm}$$

The torque is $\tau_2 = 104$ Nm.

Example 182. As illustrated below, a 60-kg monkey hangs from one end of a 12.0-m long rod, while an 80-kg monkey stands 4.0 m from the opposite end. A fulcrum rests beneath the center of the rod.

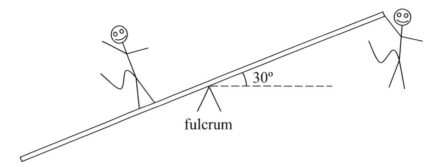

(A) What torque is exerted by the hanging monkey?

Solution. Draw a picture showing $\vec{r}_1$ and $\vec{F}_1$. The specified force is the weight ($m_1\vec{g}$) of the hanging monkey, while $\vec{r}_1$ extends from the fulcrum to the hanging monkey.

Identify the quantities involved in the torque equation.

- $\vec{r}_1$ extends from the fulcrum to the hanging monkey.

$$r_1 = \frac{L}{2} = \frac{12}{2} = 6.0 \text{ m}$$

- The force is the weight of the hanging monkey: $F_1 = m_1 g$. The mass of the hanging monkey equals $m_1 = 60$ kg. The weight of the hanging monkey equals $F_1 = m_1 g = (60)(9.81) = 589$ N.
- The angle between $\vec{r}_1$ and $\vec{F}_1$ is $\theta_1 = 90° + 30° = 120°$.

Plug this information into the equation for the magnitude of the torque.

$$\tau_1 = r_1 F_1 \sin\theta_1 = (6)(589)\sin 120° = (3534)\left(\frac{\sqrt{3}}{2}\right) = 3061 \text{ Nm} = 3.06 \times 10^3 \text{ Nm}$$

The torque is $\tau_1 = 3061$ Nm $= 3.06 \times 10^3$ Nm. (If you round gravity to ≈ 10 m/s^2, then the torque is approximately $\tau_1 \approx 1800\sqrt{3}$ Nm.)

(B) What torque is exerted by the standing monkey?

Solution. Draw a picture showing $\vec{r}_2$ and $\vec{F}_2$. The specified force is the weight ($m_2\vec{g}$) of the standing monkey, while $\vec{r}_2$ extends from the fulcrum to the standing monkey.

Identify the quantities involved in the torque equation.

- $\vec{r}_2$ extends from the fulcrum to the standing monkey.

$$r_2 = \frac{L}{2} - 4 = \frac{12}{2} - 4 = 6 - 4 = 2.0 \text{ m}$$

- The force is the weight of the standing monkey: $F_2 = m_2 g$. The mass of the standing monkey equals $m_2 = 80$ kg. The weight of the standing monkey equals $F_2 = m_2 g = (80)(9.81) = 785$ N.
- The angle between $\vec{r}_2$ and $\vec{F}_2$ is $\theta_2 = 90° - 30° = 60°$.

Plug this information into the equation for the magnitude of the torque.

$$\tau_2 = -r_2 F_2 \sin\theta_2 = -(2)(785)\sin 60° = -(1570)\left(\frac{\sqrt{3}}{2}\right) = -1360 \text{ Nm} = -1.36 \times 10^3 \text{ Nm}$$

The torque is $\tau_2 = -1360$ Nm $= -1.36 \times 10^3$ Nm. (If you round gravity to ≈ 10 m/s^2, then the torque is approximately $\tau_2 \approx -800\sqrt{3}$ Nm.) The minus **sign** represents that τ_2 is **counterclockwise**, whereas τ_1 is clockwise.

Example 183. A door is 200 cm tall and 50 cm wide. One vertical side of the door is connected to a doorway via hinges. The doorknobs are located near the other end of the door, 45 cm from the hinges. The door is presently ajar (partway open). Consider the following ways that a monkey might attempt to close the door.

(A) A monkey grabs the two doorknobs (one in each hand) and pulls with a force of 90 N directly away from the hinges. What torque does the monkey exert on the door, if any?

Solution. Identify the quantities involved in the torque equation.

- $\vec{r}_1$ extends from the hinged edge of the door to the doorknob.

$$r_1 = 45 \text{ cm} = 0.45 \text{ m} = \frac{45}{100} \text{ m} = \frac{9}{20} \text{ m}$$

- The force is the monkey's pull: $F_1 = 90$ N.

- $\theta_1 = 0°$ since $\vec{r}_1$ and $\vec{F}_1$ are **parallel** (both are directed away from the hinges).

Plug this information into the equation for the magnitude of the torque.

$$\tau_1 = r_1 F_1 \sin \theta_1 = (0.45)(90) \sin 0° = 0$$

The torque is $\tau_1 = 0$ (since $\sin 0° = 0$). You can't close a door by pulling it away from the hinges.

(B) A monkey pushes with a force of 60 N on one doorknob, pushing perpendicular to the plane of the door. What torque does the monkey exert on the door, if any?

Solution. Identify the quantities involved in the torque equation.

- $\vec{r}_2$ extends from the hinged edge of the door to the doorknob.

$$r_2 = 45 \text{ cm} = 0.45 \text{ m} = \frac{45}{100} \text{ m} = \frac{9}{20} \text{ m}$$

- The force is the monkey's push: $F_2 = 60$ N.

- $\theta_2 = 90°$ since $\vec{r}_2$ is perpendicular to and $\vec{F}_2$ (because $\vec{r}_2$ lies in the plane of the door while $\vec{F}_2$ is perpendicular to the door).

Plug this information into the equation for the magnitude of the torque.

$$\tau_2 = r_2 F_2 \sin \theta_2 = (0.45)(60) \sin 90° = 27 \text{ Nm}$$

The torque is $\tau_2 = 27$ Nm.

(C) A monkey pushes on the geometric center of the door with a force of 80 N, pushing in a direction 30° from the normal (that is, the line perpendicular to the plane of the door). What torque does the monkey exert on the door, if any?

Solution. Identify the quantities involved in the torque equation.

- $\vec{r}_3$ extends from the hinged edge of the door to the geometric **center** of the door.

$$r_3 = \frac{W}{2} = \frac{50}{2} \text{ cm} = 25 \text{ cm} = \frac{1}{4} \text{ m}$$

- The force is the monkey's push: $F_3 = 80$ N.
- Since $\vec{F}_3$ makes an angle of 30° with the normal, it makes either an angle of $\theta_3 = 60°$ or 120° with the plane of the door (while $\vec{r}_3$ lies in the plane of the door). It won't matter whether θ_3 is 60° or 120° since $\sin 60° = \sin 120°$.

Plug this information into the equation for the magnitude of the torque.

$$\tau_3 = r_3 F_3 \sin \theta_3 = \left(\frac{1}{4}\right)(80)\left(\frac{\sqrt{3}}{2}\right) = 10\sqrt{3} \text{ Nm} = 17 \text{ Nm}$$

The torque is $\tau_2 = 10\sqrt{3}$ Nm, which works out to 17 Nm.

Example 184. The force $\vec{F} = 3\,\hat{x} + 2\,\hat{y} + 4\,\hat{z}$ exerts a torque for which $\vec{r} = 6\,\hat{x} + \hat{y} - 5\hat{z}$. Find the torque vector.

Solution. Torque is the **vector product** (Chapter 29) between the position vector ($\vec{r}$) and the force vector ($\vec{F}$). Identify the components of $\vec{r}$ and $\vec{F}$.

- Compare $\vec{r} = 6\,\hat{x} + \hat{y} - 5\hat{z}$ to $\vec{r} = x\hat{x} + y\hat{y} + z\hat{z}$. The components of the position vector $\vec{r}$ are $x = 6$, $y = 1$, and $z = -5$.

- Compare $\vec{F} = 3\,\hat{x} + 2\,\hat{y} + 4\,\hat{z}$ to $\vec{F} = F_x\hat{x} + F_y\hat{y} + F_z\hat{z}$. The components of the force vector $\vec{F}$ are $F_x = 3$, $F_y = 2$, and $F_z = 4$.

Use the determinant form of the vector product.

$$\vec{\tau} = \vec{r} \times \vec{F} = \begin{vmatrix} \hat{x} & \hat{y} & \hat{z} \\ x & y & z \\ F_x & F_y & F_z \end{vmatrix} = \begin{vmatrix} \hat{x} & \hat{y} & \hat{z} \\ 6 & 1 & -5 \\ 3 & 2 & 4 \end{vmatrix}$$

$$\vec{\tau} = \hat{x}\begin{vmatrix} 1 & -5 \\ 2 & 4 \end{vmatrix} - \hat{y}\begin{vmatrix} 6 & -5 \\ 3 & 4 \end{vmatrix} + \hat{z}\begin{vmatrix} 6 & 1 \\ 3 & 2 \end{vmatrix}$$

$$\vec{\tau} = \hat{x}[(1)(4) - (-5)(2)] - \hat{y}[(6)(4) - (-5)(3)] + \hat{z}[(6)(2) - (1)(3)]$$

$$\vec{\tau} = \hat{x}[4 - (-10)] - \hat{y}[24 - (-15)] + \hat{z}(12 - 3) = \hat{x}(4 + 10) - \hat{y}(24 + 15) + 9\,\hat{z}$$

$$\vec{\tau} = 14\,\hat{x} - 39\,\hat{y} + 9\,\hat{z}$$

The torque is $\vec{\tau} = 14\,\hat{x} - 39\,\hat{y} + 9\,\hat{z}$.

31 STATIC EQUILIBRIUM

Static Equilibrium
$$\sum F_x = 0 \quad , \quad \sum F_y = 0 \quad , \quad \sum \tau = 0$$
Torque
$$\tau = r\,F\sin\theta$$
Magnitude and Direction of the Hingepin Force
$$F_H = \sqrt{H^2 + V^2} \quad , \quad \theta_H = \tan^{-1}\left(\frac{V}{H}\right)$$
Horizontal and Vertical Components of the Hingepin Force
$$H = F_H \cos\theta_H \quad , \quad V = F_H \sin\theta_H$$

Symbol	Quantity	Units
r	distance from the axis of rotation to the point where the force is applied	m
F	force	N
θ	the angle between $\vec{r}$ and $\vec{F}$	° or rad
τ	torque	Nm
F_H	magnitude of the hingepin force	N
θ_H	direction of the hingepin force	° or rad
H	horizontal component of the hingepin force	N
V	vertical component of the hingepin force	N

Example 185. In the diagram below, a 60-kg black box rests on a plank, 5.0 m to the left of the fulcrum. The fulcrum rests beneath the center of the plank. Where should a 75-kg white box be placed in order for the system to be in static equilibrium?

Solution. Begin by drawing an extended free-body diagram (FBD) for the plank. Draw each force where it effectively acts on the plank.

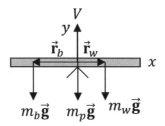

Consider each force and the torque that it exerts on the plank. We choose clockwise to be the positive sense of rotation, such that counterclockwise is negative.

- The weight of the black box ($m_b\vec{\mathbf{g}}$) pulls downward a distance of $r_b = 5.0$ m to the left of the fulcrum. Since $m_b\vec{\mathbf{g}}$ is vertical (down) and $\vec{\mathbf{r}}_b$ is horizontal (left), the angle is $\theta_b = 90°$. The resulting torque is **counterclockwise**, so it is negative.

$$\tau_b = -r_b m_b g \sin \theta_b = -(5)(60)(9.81) \sin 90°$$

- The weight of the white box ($m_w\vec{\mathbf{g}}$) pulls downward an unknown distance of r_w to the right of the fulcrum. Since $m_w\vec{\mathbf{g}}$ is vertical (down) and $\vec{\mathbf{r}}_w$ is horizontal (right), the angle is $\theta_w = 90°$. The resulting torque is **clockwise**, so it is positive.

$$\tau_w = r_w m_w g \sin \theta_w = r_w(75)(9.81) \sin 90°$$

- The weight of the plank ($m_p\vec{\mathbf{g}}$) effectively acts at the center of the plank (that's where gravity acts on average), directly above the fulcrum. Therefore, $r_p = 0$ and $\tau_p = 0$. The plank's own weight doesn't result in a torque.

- The fulcrum exerts an upward support force ($\vec{\mathbf{V}}$). Since this force acts at the fulcrum, $r_f = 0$ and $\tau_f = 0$. The fulcrum's support force doesn't result in a torque.

In static equilibrium, the net torque equals zero.

$$\sum \tau = \tau_b + \tau_w + \tau_p + \tau_f = 0$$
$$-r_b m_b g \sin \theta_b + r_w m_w g \sin \theta_w + 0 + 0 = 0$$
$$-(5)(60)(9.81) \sin 90° + r_w(75)(9.81) \sin 90° = 0$$

Divide both sides by 9.81 (g cancels out). Note that $\sin 90° = 1$. Isolate the unknown.

$$-(5)(60) + r_w(75) = 0$$
$$75 r_w = (5)(60)$$
$$r_w = \frac{(5)(6)}{75} = \frac{300}{75} = 4.0 \text{ m}$$

The white box should be placed $r_w = 4.0$ m to the right of the fulcrum.

Example 186. In the diagram below, a 20-kg white box rests on a plank, 2.0 m to the right of the fulcrum, while a black box rests 5.0 m to the left of the fulcrum. The fulcrum rests beneath the center of the plank. The system is in static equilibrium. Find the mass of the black box.

Solution. Begin by drawing an extended free-body diagram (FBD) for the plank. Draw each force where it effectively acts on the plank.

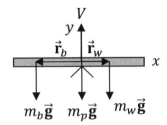

Consider each force and the torque that it exerts on the plank. We choose clockwise to be the positive sense of rotation, such that counterclockwise is negative.

- The weight of the white box ($m_w\vec{\mathbf{g}}$) pulls downward a distance of $r_w = 2.0$ m to the right of the fulcrum. Since $m_w\vec{\mathbf{g}}$ is vertical (down) and $\vec{\mathbf{r}}_w$ is horizontal (right), the angle is $\theta_w = 90°$. The resulting torque is **clockwise**, so it is positive.
$$\tau_w = r_w m_w g \sin\theta_w = (2)(20)(9.81)\sin 90°$$
- The weight of the black box ($m_b\vec{\mathbf{g}}$) pulls downward a distance of $r_b = 5.0$ m to the left of the fulcrum. Since $m_b\vec{\mathbf{g}}$ is vertical (down) and $\vec{\mathbf{r}}_b$ is horizontal (left), the angle is $\theta_b = 90°$. The resulting torque is **counterclockwise**, so it is negative.
$$\tau_b = -r_b m_b g \sin\theta_b = -(5)m_b(9.81)\sin 90°$$
- The weight of the plank ($m_p\vec{\mathbf{g}}$) effectively acts at the center of the plank (where gravity acts on average), directly above the fulcrum. Therefore, $r_p = 0$ and $\tau_p = 0$.
- The fulcrum exerts an upward support force ($\vec{\mathbf{V}}$). Since this force acts at the fulcrum, $r_f = 0$ and $\tau_f = 0$.

In static equilibrium, the net torque equals zero.
$$\sum \tau = \tau_b + \tau_w + \tau_p + \tau_f = 0$$
$$-r_b m_b g \sin\theta_b + r_w m_w g \sin\theta_w + 0 + 0 = 0$$
$$-(5)m_b(9.81)\sin 90° + (2)(20)(9.81)\sin 90° = 0$$
Divide both sides by 9.81 (g cancels out). Note that $\sin 90° = 1$. Isolate the unknown.
$$-(5)m_b + (2)(20) = 0$$
$$(2)(20) = 5m_b$$
$$m_b = \frac{(2)(20)}{5} = \frac{40}{5} = 8.0 \text{ kg}$$
The mass of the black box is $m_b = 8.0$ kg.

Example 187. In the diagram below, a 100-kg black box rests on a plank, 4.0 m to the left of the fulcrum, and a 50-kg white box rests 16.0 m to the left of the fulcrum. The fulcrum rests beneath the center of the plank. Where should a 60-kg gray box be placed in order for the system to be in static equilibrium?

Solution. Begin by drawing an extended free-body diagram (FBD) for the plank. Draw each force where it effectively acts on the plank.

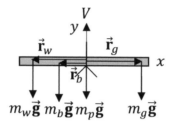

Consider each force and the torque that it exerts on the plank. We choose clockwise to be the positive sense of rotation, such that counterclockwise is negative.

- The weight of the black box ($m_b\vec{\mathbf{g}}$) pulls downward a distance of $r_b = 4.0$ m to the left of the fulcrum. The resulting torque is **counterclockwise**, so it is negative.
$$\tau_b = -r_b m_b g \sin\theta_b = -(4)(100)(9.81)\sin 90°$$
- The weight of the white box ($m_w\vec{\mathbf{g}}$) pulls downward a distance of $r_w = 16.0$ m to the left of the fulcrum. The resulting torque is **counterclockwise**, so it is negative.
$$\tau_w = -r_w m_w g \sin\theta_w = -(16)(50)(9.81)\sin 90°$$
- The weight of the gray box ($m_g\vec{\mathbf{g}}$) pulls downward an unknown distance of r_g to the right of the fulcrum. The resulting torque is **clockwise**, so it is positive.
$$\tau_g = r_g m_g g \sin\theta_g = r_g(60)(9.81)\sin 90°$$
- The weight of the plank ($m_p\vec{\mathbf{g}}$) and the upward support force of the fulcrum ($\vec{\mathbf{V}}$) don't affect the rotation of the plank: $\tau_p = 0$ and $\tau_f = 0$.

In static equilibrium, the net torque equals zero.
$$\sum \tau = \tau_b + \tau_w + \tau_g + \tau_p + \tau_f = 0$$
$$-r_b m_b g \sin\theta_b - r_w m_w g \sin\theta_w + r_g m_g g \sin\theta_g + 0 + 0 = 0$$
$$-(4)(100)(9.81)\sin 90° - (16)(50)(9.81)\sin 90° + r_g(60)(9.81)\sin 90° = 0$$
Divide both sides by 9.81 (g cancels out). Note that $\sin 90° = 1$. Isolate the unknown.
$$-(4)(100) - (16)(50) + r_g(60) = 0$$
$$60 r_g = (4)(100) + (16)(50)$$
$$r_g = \frac{(4)(100) + (16)(50)}{60} = \frac{400 + 800}{60} = \frac{1200}{60} = 20.0 \text{ m}$$
The gray box should be placed $r_g = 20.0$ m to the right of the fulcrum.

Example 188. In the diagram below, a 30-kg white box rests on a plank, 2.5 m to the left of the fulcrum, and a 40-kg black box rests 7.5 m to the right of the fulcrum. The fulcrum rests beneath the center of the plank. Where should a 25-kg gray box be placed in order for the system to be in static equilibrium?

Solution. Begin by drawing an extended free-body diagram (FBD) for the plank. Draw each force where it effectively acts on the plank.

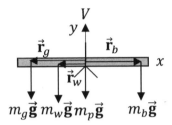

Consider each force and the torque that it exerts on the plank. We choose clockwise to be the positive sense of rotation, such that counterclockwise is negative.

- The weight of the white box ($m_w\vec{\mathbf{g}}$) pulls downward a distance of $r_w = 2.5$ m to the left of the fulcrum. The resulting torque is **counterclockwise**, so it is negative.
$$\tau_w = -r_w m_w g \sin\theta_w = -(2.5)(30)(9.81)\sin 90°$$
- The weight of the black box ($m_b\vec{\mathbf{g}}$) pulls downward a distance of $r_b = 7.5$ m to the left of the fulcrum. The resulting torque is **clockwise**, so it is positive.
$$\tau_b = r_b m_b g \sin\theta_b = (7.5)(40)(9.81)\sin 90°$$
- The weight of the gray box ($m_g\vec{\mathbf{g}}$) pulls downward an unknown distance of r_g to the left of the fulcrum. The resulting torque is **counterclockwise**, so it is negative.
$$\tau_g = -r_g m_g g \sin\theta_g = -r_g(25)(9.81)\sin 90°$$
- The weight of the plank ($m_p\vec{\mathbf{g}}$) and the upward support force of the fulcrum ($\vec{\mathbf{V}}$) don't affect the rotation of the plank: $\tau_p = 0$ and $\tau_f = 0$.

In static equilibrium, the net torque equals zero.
$$\sum\tau = \tau_w + \tau_b + \tau_g + \tau_p + \tau_f = 0$$
$$-r_w m_w g \sin\theta_w + r_b m_b g \sin\theta_b - r_g m_g g \sin\theta_g + 0 + 0 = 0$$
$$-(2.5)(30)(9.81)\sin 90° + (7.5)(40)(9.81)\sin 90° - r_g(25)(9.81)\sin 90° = 0$$
Divide both sides by 9.81 (g cancels out). Note that $\sin 90° = 1$. Isolate the unknown.
$$-(2.5)(30) + (7.5)(40) - r_g(25) = 0$$
$$-(2.5)(30) + (7.5)(40) = 25 r_g$$
$$r_g = \frac{-(2.5)(30) + (7.5)(40)}{25} = \frac{-75 + 300}{25} = \frac{225}{25} = 9.0 \text{ m}$$
The gray box should be placed $r_g = 9.0$ m to the left of the fulcrum.

Example 189. As illustrated below, a 40-kg monkey is suspended from a rope in static equilibrium. The monkey's rope is connected to two other ropes which are supported from the ceiling. Determine the tension in each cord.

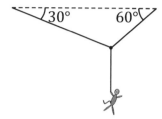

Solution. The trick to this problem is to draw a FBD for the **knot** (where the three cords meet). Since the system is in static equilibrium, it's not rotating, so we are free to choose the axis of rotation anywhere we like. Let's choose the axis of rotation to pass through the knot, perpendicular to the page. With this choice, all of the torques are zero since each force pulls directly away from the knot (it's like pushing on the hinges to try to open a door). The sum of the torques simply states that zero equals zero in this problem. For this problem, it is useful to sum the forces (that is, apply Newton's second law).

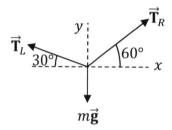

Set the sums of the components of the forces equal to zero, since the system is in static equilibrium (meaning that the acceleration is zero). Resolve the tensions into components.

$$\sum F_x = 0 \quad , \quad \sum F_y = 0$$
$$T_R \cos 60° - T_L \cos 30° = 0 \quad , \quad T_R \sin 60° + T_L \sin 30° - mg = 0$$

Solve for T_R in terms of T_L in the x-sum.

$$T_R \cos 60° = T_L \cos 30°$$
$$T_R = \frac{T_L \cos 30°}{\cos 60°} = T_L \frac{\sqrt{3}/2}{1/2} = T_L \frac{\sqrt{3}}{2}\frac{2}{1} = T_L \sqrt{3}$$

To divide by a fraction, multiply by its **reciprocal**. Note that the reciprocal of $\frac{1}{2}$ is 2. Plug this expression in for T_R in the y-sum.

$$T_R \sin 60° + T_L \sin 30° - mg = 0$$
$$\left(T_L \sqrt{3}\right) \sin 60° + T_L \sin 30° - mg = 0$$
$$T_L \sqrt{3}\left(\frac{\sqrt{3}}{2}\right) + T_L \left(\frac{1}{2}\right) = mg$$
$$\frac{T_L \sqrt{3}\sqrt{3}}{2} + \frac{T_L}{2} = mg$$

Note that $\sqrt{3}\sqrt{3} = 3$.

$$\frac{3T_L}{2} + \frac{T_L}{2} = mg$$

Factor out the T_L.

$$T_L\left(\frac{3}{2} + \frac{1}{2}\right) = mg$$

$$T_L\left(\frac{3+1}{2}\right) = mg$$

$$T_L\left(\frac{4}{2}\right) = mg$$

$$2T_L = mg$$

$$T_L = \frac{mg}{2} = \frac{(40)(9.81)}{2} = 196 \text{ N}$$

Plug this value for T_L into the previous equation for T_R (from the x-sum).

$$T_R = T_L\sqrt{3} = 196\sqrt{3} = 339 \text{ N}$$

The answers are $T_L = 196 \text{ N} \approx 200 \text{ N}$ and $T_R = 339 \text{ N} \approx 200\sqrt{3} \text{ N}$. The tension in the bottom cord equals the weight of the monkey: $T_B = mg = (40)(9.81) = 392 \text{ N} \approx 400 \text{ N}$.

Example 190. As illustrated below, a uniform boom is supported by a tie rope that connects to a wall. The system is in static equilibrium. The boom has a mass of 50 kg and is 10-m long. The lower end of the boom is connected to a hingepin. The tie rope is perpendicular to the boom and can sustain a maximum tension of 600 N.

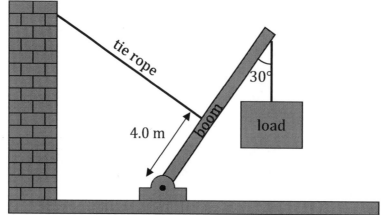

(A) What maximum load can the boom support without snapping the tie rope?

Solution. Begin by drawing an extended free-body diagram (FBD) for the boom. Draw each force where it effectively acts on the boom.

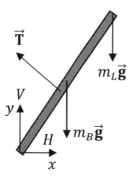

Consider each force and the torque that it exerts on the boom. We choose clockwise to be the positive sense of rotation, such that counterclockwise is negative.

- The weight of the load ($m_L\vec{\mathbf{g}}$) pulls downward a distance of $r_L = 10$ m from the hinge. The resulting torque is **clockwise**, so it is positive.

$$\tau_L = r_L m_L g \sin\theta_L = (10)m_L(9.81)\sin 150°$$

- The weight of the boom ($m_B\vec{\mathbf{g}}$) pulls downward a distance of $r_B = \frac{L}{2} = \frac{10}{2} = 5$ m from the hinge. The resulting torque is **clockwise**, so it is positive.
$$\tau_B = r_B m_B g \sin \theta_B = (5)(50)(9.81) \sin 150°$$

- The tension ($\vec{\mathbf{T}}$) pulls perpendicular to the boom a distance of $r_T = 4$ m from the hinge. The resulting torque is **counterclockwise**, so it is negative. To find the maximum load, we will use the maximum tension, which is 600 N.
$$\tau_T = -r_T T \sin \theta_T = -(4)(600) \sin 90°$$

- The horizontal (H) and vertical (V) components of the hingepin force don't affect the rotation of the boom since they pass through the hinge (which makes $r_H = 0$ and $r_V = 0$): $\tau_H = 0$ and $\tau_V = 0$.

In static equilibrium, the net torque equals zero.
$$\sum \tau = \tau_L + \tau_B + \tau_T + \tau_H + \tau_V = 0$$
$$r_L m_L g \sin \theta_L + r_B m_B g \sin \theta_B - r_T T \sin \theta_T + 0 + 0 = 0$$
$$(10)m_L(9.81) \sin 150° + (5)(50)(9.81) \sin 150° - (4)(600) \sin 90° = 0$$

Note that $\sin 90° = 1$ and $\sin 150° = \frac{1}{2}$. Isolate the unknown.
$$(10)m_L(9.81)\left(\frac{1}{2}\right) + (5)(50)(9.81)\left(\frac{1}{2}\right) - (4)(600)(1) = 0$$
$$49.1 m_L + 1226 - 2400 = 0$$
$$m_L = \frac{2400 - 1226}{49.1} = \frac{1174}{49.1} = 24 \text{ kg}$$

The maximum load is $m_L = 24$ kg.

(B) Find the maximum horizontal and vertical components of the force exerted on the hingepin.

Solution. Set the sums of the components of the forces (in Newton's second law) equal to zero, since the system is in static equilibrium (meaning that the acceleration is zero). Study the extended free-body diagram (FBD) that we drew in the beginning of our solution to part (A). Resolve the tension into components: The tension ($\vec{T}$) is directed $30°$ above the negative x-axis, such that $-T\cos 30°$ and $T\sin 30°$ appear in the x- and y-sums.

$$\sum F_x = 0 \quad , \quad \sum F_y = 0$$

$$H - T\cos 30° = 0 \quad , \quad V + T\sin 30° - m_B g - m_L g = 0$$

$$H = T\cos 30° \quad , \quad V = m_B g + m_L g - T\sin 30°$$

$$H = (600)\left(\frac{\sqrt{3}}{2}\right) \quad , \quad V = (50)(9.81) + (24)(9.81) - (600)\left(\frac{1}{2}\right)$$

$$H = 300\sqrt{3}\,\text{N} = 520\,\text{N} \quad , \quad V = 426\,\text{N}$$

The horizontal and vertical components of the hingepin force are $H = 300\sqrt{3}\,\text{N} = 520\,\text{N}$ and $V = 426\,\text{N}$.

32 MOMENT OF INERTIA

Pointlike Object	System of Objects
$I = mR^2$	$I = I_1 + I_2 + \cdots + I_N$
Parallel-axis Theorem	**Continuous Distribution of Mass**
$I_p = I_{CM} + mh^2$	$I = \int r_\perp^2 \, dm$

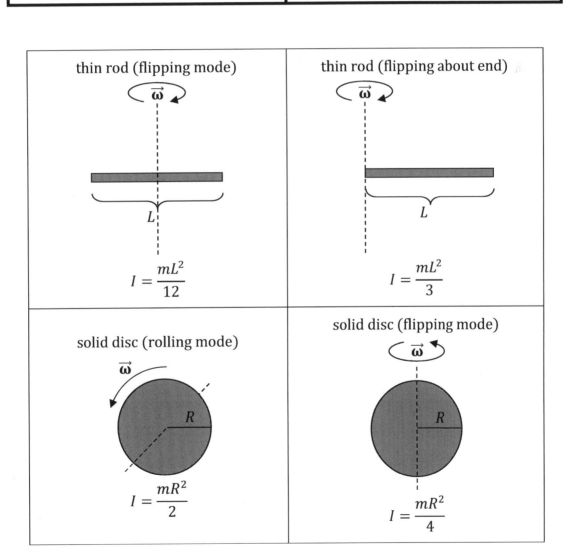

thin rod (flipping mode)

$$I = \frac{mL^2}{12}$$

thin rod (flipping about end)

$$I = \frac{mL^2}{3}$$

solid disc (rolling mode)

$$I = \frac{mR^2}{2}$$

solid disc (flipping mode)

$$I = \frac{mR^2}{4}$$

thin ring (rolling mode)

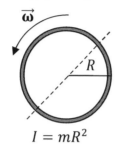

$$I = mR^2$$

thin ring (flipping mode)

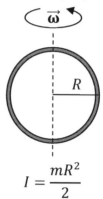

$$I = \frac{mR^2}{2}$$

thick ring (rolling mode)

$$I = m\frac{R_0^2 + R^2}{2}$$

thick ring (flipping mode)

$$I = m\frac{R_0^2 + R^2}{4}$$

solid cylinder (rolling mode)

$$I = \frac{mR^2}{2}$$

thin hollow cylinder (rolling mode)

$$I = mR^2$$

thick hollow cylinder (rolling mode)

$$I = m \frac{R_0^2 + R^2}{2}$$

solid sphere (rolling mode)

$$I = \frac{2mR^2}{5}$$

thin hollow sphere (rolling mode)

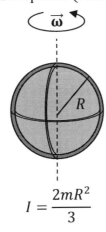

$$I = \frac{2mR^2}{3}$$

thick hollow sphere (rolling mode)

$$I = \frac{2m}{5} \frac{R^5 - R_0^5}{R^3 - R_0^3}$$

solid cube about an axis through its center and perpendicular to a face

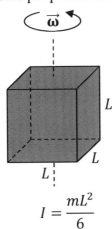

$$I = \frac{mL^2}{6}$$

solid rectangle about an axis through its center and perpendicular to the rectangle

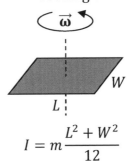

$$I = m \frac{L^2 + W^2}{12}$$

solid rectangle (flipping mode)

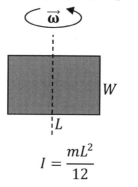

$$I = \frac{mL^2}{12}$$

solid cone about its symmetry axis

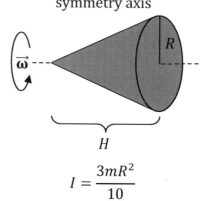

$$I = \frac{3mR^2}{10}$$

solid single-holed ring torus (rolling mode)

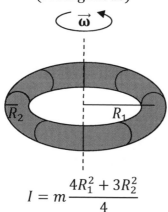

$$I = m\frac{4R_1^2 + 3R_2^2}{4}$$

solid ellipse about a symmetric bisector

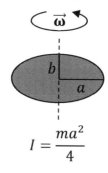

$$I = \frac{ma^2}{4}$$

Symbol	Name	SI Units
m	mass	kg
R	radius	m
R_0	inner radius	m
L	length	m
W	width	m
I	moment of inertia	$kg{\cdot}m^2$
I_{CM}	moment of inertia about an axis passing through the center of mass	$kg{\cdot}m^2$
I_p	moment of inertia about an axis parallel to the axis used for I_{CM}	$kg{\cdot}m^2$
h	distance between the two parallel axes used for I_{CM} and I_p	m

Example 191. As illustrated below, three 6.0-kg masses are joined together with approximately massless rods in the shape of an equilateral triangle with an edge length of $L = 4.0$ m. The 6.0-kg masses are small in size compared to the length of the rods.

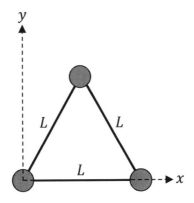

(A) Determine the moment of inertia of the system about the x-axis.

Solution. Since the rods are very light (approximately massless compared to the 6.0-kg masses), we can neglect them. Since the 6.0-kg masses are small compared to the length of the rods, we can treat them as pointlike objects. Use the formula for the moment of inertia of three pointlike objects sharing a common axis of rotation (the x-axis).

$$I = m_1 R_1^2 + m_2 R_2^2 + m_3 R_3^2$$

All of the masses are the same ($m = 6.0$ kg). Each R is the distance of the corresponding mass from the axis of rotation (the x-axis).

- The left and right masses lie on the x-axis. Therefore, $R_1 = R_3 = 0$.
- For the top mass, R_2 is the height of the triangle. Split the triangle in two (as shown below) to make a right triangle. The hypotenuse is $L = 4.0$ m and the base is $\frac{L}{2} = \frac{4}{2} = 2.0$ m. Use the Pythagorean theorem to determine the height of the triangle.

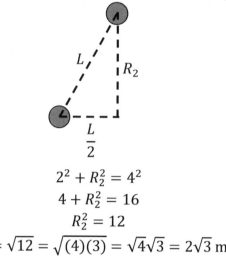

$$2^2 + R_2^2 = 4^2$$
$$4 + R_2^2 = 16$$
$$R_2^2 = 12$$
$$R_2 = \sqrt{12} = \sqrt{(4)(3)} = \sqrt{4}\sqrt{3} = 2\sqrt{3} \text{ m}$$

Plug these values into the previous equation for moment of inertia.

$$I = (6)(0)^2 + (6)\left(2\sqrt{3}\right)^2 + (6)(0)^2 = 0 + (6)(12) + 0 = 72 \text{ kg·m}^2$$

The moment of inertia about the x-axis is $I = 72$ kg·m².

(B) Determine the moment of inertia of the system about the y-axis.

Solution. Solve this problem the same was as in part (A), except that now the R's will be different because the axis of rotation is the y-axis (instead of the x-axis). Each R is the distance of the corresponding mass from the y-axis.

- The left mass lies on the y-axis. Therefore, $R_1 = 0$.
- The top mass lies $R_2 = \frac{L}{2} = \frac{4}{2} = 2.0$ m from the y-axis.
- The right mass lies $R_3 = L = 4.0$ m from the y-axis.

Use the equation for the moment of inertia of a system of pointlike objects.

$$I = m_1 R_1^2 + m_2 R_2^2 + m_3 R_3^2 = (6)(0)^2 + (6)(2)^2 + (6)(4)^2$$
$$I = 0 + 24 + 96 = 120 \text{ kg·m}^2$$

The moment of inertia about the y-axis is $I = 120$ kg·m².

(C) Determine the moment of inertia of the system about the z-axis.

Solution. Note that the z-axis is perpendicular to the page and passes through the origin. Solve this problem the same was as in part (A), except that now the R's will be different because the axis of rotation is the z-axis (instead of the x-axis). Each R is the distance of the corresponding mass from the z-axis. Since all three masses lie in the xy plane, each R is the distance of the corresponding mass to the **origin**.

- The left mass lies on the origin. Therefore, $R_1 = 0$.
- The top and right masses lie $R_2 = R_3 = L = 4.0$ m from the origin.

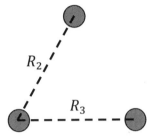

Use the equation for the moment of inertia of a system of pointlike objects.
$$I = m_1 R_1^2 + m_2 R_2^2 + m_3 R_3^2 = (6)(0)^2 + (6)(4)^2 + (6)(4)^2$$
$$I = 0 + 96 + 96 = 192 \text{ kg·m}^2$$
The moment of inertia about the z-axis is $I = 192$ kg·m². **Check:** Note that $I_x + I_y = I_z$. This is true for a planar object lying in the xy plane. (It's called the perpendicular axis theorem, not to be confused with the parallel-axis theorem that we will encounter in examples later in this chapter.)

Example 192. Three bananas are joined together with approximately massless rods. The bananas are small in size compared to the length of the rods. The masses and the (x, y) coordinates of the bananas are listed below.

- A 400-g banana has coordinates $(3.0$ m, $0)$.
- A 300-g banana has coordinates $(0, 2.0$ m$)$.
- A 500-g banana has coordinates $(-4.0$ m, 1.0 m$)$.

(A) Determine the moment of inertia of the system about the x-axis.

Solution. List the masses and R's for each banana, where each R is the distance of the corresponding banana from the axis of rotation (the x-axis). Note that the distance of each banana from the x-axis is its y-coordinate.

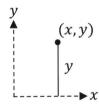

- $m_1 = 400$ g $= 0.4$ kg $= \frac{2}{5}$ kg and $R_1 = y_1 = 0$.

- $m_2 = 300 \text{ g} = 0.3 \text{ kg} = \frac{3}{10} \text{ kg and } R_2 = y_2 = 2.0 \text{ m}.$
- $m_3 = 500 \text{ g} = 0.5 \text{ kg} = \frac{1}{2} \text{ kg and } R_3 = y_3 = 1.0 \text{ m}.$

Use the equation for the moment of inertia of a system of pointlike objects.

$$I = m_1 R_1^2 + m_2 R_2^2 + m_3 R_3^2 = \left(\frac{2}{5}\right)(0)^2 + \left(\frac{3}{10}\right)(2)^2 + \left(\frac{1}{2}\right)(1)^2$$

$$I = 0 + \frac{12}{10} + \frac{1}{2} = \frac{12}{10} + \frac{5}{10} = \frac{17}{10} \text{ kg·m}^2 = 1.7 \text{ kg·m}^2$$

To add fractions, find a **common denominator**. Note that $\frac{1}{2} = \frac{5}{10}$. The moment of inertia about the x-axis is $I = \frac{17}{10} \text{ kg·m}^2 = 1.7 \text{ kg·m}^2$.

(B) Determine the moment of inertia of the system about the y-axis.

Solution. List the masses and R's for each banana, where each R is the distance of the corresponding banana from the axis of rotation (the y-axis). Note that the distance of each banana from the y-axis is its x-coordinate.

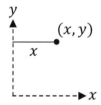

- $m_1 = 400 \text{ g} = 0.4 \text{ kg} = \frac{2}{5} \text{ kg and } R_1 = x_1 = 3.0 \text{ m}.$
- $m_2 = 300 \text{ g} = 0.3 \text{ kg} = \frac{3}{10} \text{ kg and } R_2 = x_2 = 0.$
- $m_3 = 500 \text{ g} = 0.5 \text{ kg} = \frac{1}{2} \text{ kg and } R_3 = x_3 = -4.0 \text{ m}.$

Use the equation for the moment of inertia of a system of pointlike objects.

$$I = m_1 R_1^2 + m_2 R_2^2 + m_3 R_3^2 = \left(\frac{2}{5}\right)(3)^2 + \left(\frac{3}{10}\right)(0)^2 + \left(\frac{1}{2}\right)(-4)^2$$

$$I = \frac{18}{5} + 0 + 8 = \frac{18}{5} + \frac{40}{5} = \frac{58}{5} \text{ kg·m}^2 = 11.6 \text{ kg·m}^2$$

To add fractions, find a **common denominator**. Note that $8 = \frac{40}{5}$. Also note that $(-4)^2 = +16$. The moment of inertia about the y-axis is $I = \frac{58}{5} \text{ kg·m}^2 = 11.6 \text{ kg·m}^2$.

(C) Determine the moment of inertia of the system about the z-axis.

Solution. Note that the z-axis is perpendicular to the page and passes through the origin. Solve this problem the same was as in part (A), except that now the R's will be different because the axis of rotation is the z-axis (instead of the x-axis). Each R is the distance of the corresponding mass from the z-axis. Since all three masses lie in the xy plane, each R is the distance of the corresponding mass to the **origin**. Apply the Pythagorean theorem to determine R from the (x, y) coordinates.

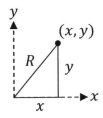

- $m_1 = 400 \text{ g} = 0.4 \text{ kg} = \frac{2}{5} \text{ kg}$ and $R_1 = \sqrt{x_1^2 + y_1^2} = \sqrt{3^2 + 0^2} = 3.0 \text{ m}$.

- $m_2 = 300 \text{ g} = 0.3 \text{ kg} = \frac{3}{10} \text{ kg}$ and $R_2 = \sqrt{x_2^2 + y_2^2} = \sqrt{0^2 + 2^2} = 2.0 \text{ m}$.

- $m_3 = 500 \text{ g} = 0.5 \text{ kg} = \frac{1}{2} \text{ kg}$ and $R_3 = \sqrt{x_3^2 + y_3^2} = \sqrt{(-4)^2 + 1^2} = \sqrt{17} \text{ m}$.

Use the equation for the moment of inertia of a system of pointlike objects.

$$I = m_1 R_1^2 + m_2 R_2^2 + m_3 R_3^2 = \left(\frac{2}{5}\right)(3)^2 + \left(\frac{3}{10}\right)(2)^2 + \left(\frac{1}{2}\right)\left(\sqrt{17}\right)^2$$

$$I = \frac{18}{5} + \frac{12}{10} + \frac{17}{2} = \frac{36}{10} + \frac{12}{10} + \frac{85}{10} = \frac{133}{10} \text{ kg·m}^2 = 13.3 \text{ kg·m}^2$$

To add fractions, find a **common denominator**. Note that $\frac{18}{5} = \frac{36}{10}$ and $\frac{17}{2} = \frac{85}{10}$. The moment of inertia about the y-axis is $I = \frac{133}{10} \text{ kg·m}^2 = 13.3 \text{ kg·m}^2$. **Check**: Note that $I_x + I_y = I_z$, since $\frac{17}{10} + \frac{58}{5} = \frac{17+116}{10} = \frac{133}{10}$. This is true for a planar object lying in the xy plane. (It's called the perpendicular axis theorem, not to be confused with the parallel-axis theorem that we will encounter in examples later in this chapter.)

Example 193. The diagram below shows a uniform rod that is welded onto two hollow spheres. (The ends of the rod are joined to each sphere. The rod doesn't penetrate inside either sphere, that is no holes are drilled through the spheres.) The rod has a mass of 6.0 kg and length of 8.0 m, the left sphere has a mass of 9.0 kg and a radius of 2.0 m, and the right sphere has a mass of 3.0 kg and a radius of 1.0 m. Determine the moment of inertia about the axis shown, which bisects the rod.

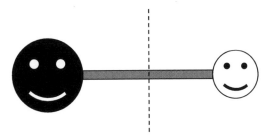

Solution. Find the moment of inertia of each object (the left sphere, the rod, and the right sphere) and then add their moments of inertia together. Don't use the formula for a point-like object. Instead, use the formulas for a hollow sphere and for a uniform rod. We need the moments of inertia about the indicated axis of rotation (the dashed line in the figure above), which passes perpendicularly through the midpoint of the rod.

Look up the formula (in the tables at the beginning of this chapter) for the moment of inertia of a rod rotating about an axis through its center and perpendicular to the rod.

$$I_{rod} = \frac{1}{12} m_r L^2 = \frac{1}{12}(6)(8)^2 = 32 \text{ kg·m}^2$$

For the hollow spheres, we must apply the **parallel-axis theorem**.

$$I_{hs} = I_{CM} + m_{hs}h^2$$

Here, I_{CM} is the moment of inertia of a hollow sphere rotating about an axis passing through its center, while I_{hs} is the moment of inertia of a hollow sphere about an axis that is parallel to an axis passing through its center. Look up the formula (in the tables at the beginning of this chapter) for the moment of inertia of a hollow sphere about an axis passing through its center.

$$I_{CM} = \frac{2}{3} m_{hs} R^2$$

Substitute this into the equation for I_{hs}.

$$I_{hs} = I_{CM} + m_{hs}h^2 = \frac{2}{3} m_{hs} R^2 + m_{hs} h^2$$

Note that $h = \frac{L}{2} + R$, since h is the distance from the actual axis of rotation (passing through the center of the rod) and the center of the sphere.

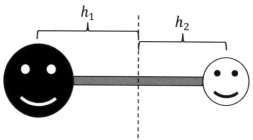

Use the radius of each sphere to determine h_1 and h_2.

$$h_1 = \frac{L}{2} + R_1 = \frac{8}{2} + 2 = 4 + 2 = 6 \text{ m}$$

$$h_2 = \frac{L}{2} + R_2 = \frac{8}{2} + 1 = 4 + 1 = 5 \text{ m}$$

Plug these values into the equation for I_{hs}.

$$I_{hs1} = \frac{2}{3} m_{hs1} R_1^2 + m_{hs1} h_1^2 = \frac{2}{3}(9)(2)^2 + (9)(6)^2 = 24 + 324 = 348 \text{ kg·m}^2$$

$$I_{hs2} = \frac{2}{3} m_{hs2} R_2^2 + m_{hs2} h_2^2 = \frac{2}{3}(3)(1)^2 + (3)(5)^2 = 2 + 75 = 77 \text{ kg·m}^2$$

Add the three moments of inertia together:

$$I = I_{hs1} + I_{rod} + I_{hs2} = 32 + 348 + 77 = 457 \text{ kg·m}^2$$

The moment of inertia of the composite object is $I = 457 \text{ kg·m}^2$.

Differential Arc Length		
$ds = dx$ (along x)	$ds = dy$ (along y)	$ds = Rd\theta$ (circular arc)

Differential Area Element		
$dA = dxdy$ (polygon)	$dA = rdrd\theta$ (pie slice)	$dA = R^2 \sin\theta \, d\theta d\varphi$ (spherical surface)

Differential Volume Element		
$dV = dxdydz$ (bounded by flat sides)	$dV = r_c dr_c d\theta dz$ (cylinder or cone)	$dV = r^2 \sin\theta \, drd\theta d\varphi$ (spherical)

Differential Mass Element		
$dm = \lambda ds$	$dm = \sigma dA$	$dm = \rho dV$

Moment of Inertia Integral
$I = \displaystyle\int r_\perp^2 \, dm$

Cartesian Coordinates

2D Polar Coordinates

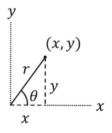

$$x = r\cos\theta \quad , \quad r = \sqrt{x^2 + y^2}$$
$$y = r\sin\theta \quad , \quad \theta = \tan^{-1}\left(\frac{y}{x}\right)$$

Cylindrical Coordinates

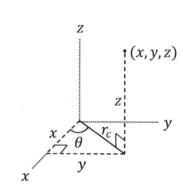

$$x = r_c\cos\theta \quad , \quad r_c = \sqrt{x^2 + y^2}$$
$$y = r_c\sin\theta \quad , \quad \theta = \tan^{-1}\left(\frac{y}{x}\right)$$

Spherical Coordinates

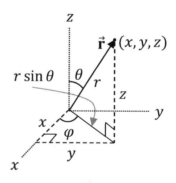

$$x = r\sin\theta\cos\varphi \quad , \quad r = \sqrt{x^2 + y^2 + z^2}$$
$$y = r\sin\theta\sin\varphi \quad , \quad \theta = \cos^{-1}\left(\frac{z}{r}\right)$$
$$z = r\cos\theta \quad , \quad \varphi = \tan^{-1}\left(\frac{y}{x}\right)$$

Symbol	Name	SI Units
dm	differential mass element	kg
m	total mass of the object	kg
$r_\perp$	shortest distance from dm to the axis of rotation	m
x, y, z	Cartesian coordinates of dm	m, m, m
r, θ	2D polar coordinates of dm	m, rad
r_c, θ, z	cylindrical coordinates of dm	m, rad, m
r, θ, φ	spherical coordinates of dm	m, rad, rad
λ	linear mass density (for an arc)	kg/m
σ	mass density for a surface area	kg/m^2
ρ	mass density for a 3D solid	kg/m^3
ds	differential arc length	m
dA	differential area element	m^2
dV	differential volume element	m^3
I	moment of inertia	$kg \cdot m^2$

Note: The symbols λ, σ, and ρ are the lowercase Greek letters lambda, sigma, and rho.

Example 194. A rod with endpoints at the origin and the point $(0, L)$ has non-uniform linear mass density $\lambda(y) = \beta\sqrt{y}$, where β is a positive constant. Derive an equation for the moment of inertia of the rod about the x-axis in terms of the total mass of the rod, m, and the length of the rod, L.

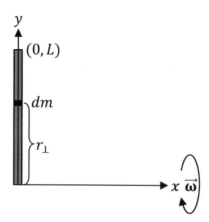

Solution. Begin with a labeled diagram. Draw a representative dm and label $r_\perp$ for that dm, as we did in the diagram above. When we perform the integration, we effectively integrate over every dm that makes up the rod. Perform the moment of inertia integral.

$$I = \int r_\perp^2 \, dm$$

For a rod, write $dm = \lambda ds$.

$$I = \int r_\perp^2 \, \lambda ds$$

Since the rod lies on the y-axis, $ds = dy$. The limits of integration are from $y = 0$ to $y = L$.

$$I = \int_{y=0}^{L} r_\perp^2 \lambda \, dy$$

Don't pull λ or $r_\perp^2$ out of the integral. Substitute $\lambda = \beta\sqrt{y}$ (this was given in the problem).

$$I = \int_{y=0}^{L} r_\perp^2 \left(\beta\sqrt{y}\right) dy$$

Note that $r_\perp$ extends from the axis of rotation to each dm. Study the diagram above to see that $r_\perp = y$ since dm is a distance y from the axis of rotation (the x-axis). Substitute $r_\perp^2 = y^2$ into the integral.

$$I = \int_{y=0}^{L} y^2 \left(\beta\sqrt{y}\right) dy$$

Apply the rule from algebra that $\sqrt{y} = y^{1/2}$. Pull the constant β out of the integral.

$$I = \beta \int_{y=0}^{L} y^2 y^{1/2} \, dy$$

Note that $y^2 y^{1/2} = y^{5/2}$ according to the rule $y^m y^n = y^{m+n}$ since $2 + \frac{1}{2} = \frac{4}{2} + \frac{1}{2} = \frac{5}{2}$.

$$I = \beta \int_{y=0}^{L} y^{5/2} \, dy$$

Now we have an integral over y which we can perform using the method from Chapter 2.

$$I = \beta \left[\frac{y^{5/2+1}}{5/2 + 1} \right]_{y=0}^{y=L}$$

To add fractions, use a **common denominator**: $\frac{5}{2} + 1 = \frac{5}{2} + \frac{2}{2} = \frac{7}{2}$. To divide by a fraction, multiply by its **reciprocal**.

$$I = \beta \left[\frac{y^{7/2}}{7/2} \right]_{y=0}^{y=L} = \beta \left[\frac{2y^{7/2}}{7} \right]_{y=0}^{y=L} = \beta \left(\frac{2L^{7/2}}{7} - \frac{2(0)^{7/2}}{7} \right) = \frac{2\beta L^{7/2}}{7}$$

In order to eliminate the constant β from our answer, we will integrate over dm using the same substitutions as before, beginning with $dm = \lambda ds$.

$$m = \int dm = \int \lambda \, ds = \int_{y=0}^{L} \beta \sqrt{y} \, dy = \beta \int_{y=0}^{L} \sqrt{y} \, dy = \beta \int_{y=0}^{L} y^{1/2} \, dy$$

$$m = \beta \left[\frac{y^{1/2+1}}{1/2 + 1} \right]_{y=0}^{y=L}$$

To add fractions, use a **common denominator**: $\frac{1}{2} + 1 = \frac{1}{2} + \frac{2}{2} = \frac{3}{2}$. To divide by a fraction, multiply by its **reciprocal**.

$$m = \beta \left[\frac{y^{3/2}}{3/2} \right]_{y=0}^{y=L} = \beta \left[\frac{2y^{3/2}}{3} \right]_{y=0}^{y=L} = \beta \left(\frac{2L^{3/2}}{3} - \frac{2(0)^{3/2}}{3} \right) = \frac{2\beta L^{3/2}}{3}$$

Solve for β. Multiply both sides of the equation by 3 and divide by $2L^{3/2}$.

$$\beta = \frac{3m}{2L^{3/2}}$$

Substitute this expression for β into the previous expression for moment of inertia.

$$I = \frac{2\beta L^{7/2}}{7} = \frac{2L^{7/2}}{7} \left(\frac{3m}{2L^{3/2}} \right) = \frac{3m}{7} \frac{L^{7/2}}{L^{3/2}} = \frac{3mL^2}{7}$$

Apply the rule $\frac{L^m}{L^n} = L^{m-n}$. Note that $\frac{7}{2} - \frac{3}{2} = \frac{4}{2} = 2$, such that $\frac{L^{7/2}}{L^{3/2}} = L^{4/2} = L^2$. The moment of inertia of the rod about the x-axis is $I = \frac{3mL^2}{7}$.

Example 195. The right triangle illustrated below has uniform density. Derive an equation for its moment of inertia about the x-axis in terms of b, h, and/or m.

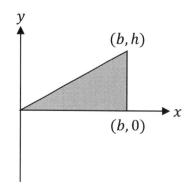

Solution. Begin with a labeled diagram. Draw a representative dm and label $r_\perp$, as we did in the diagram below. When we perform the integration, we effectively integrate over every dm that makes up the solid triangle. Each dm has coordinates (x, y).

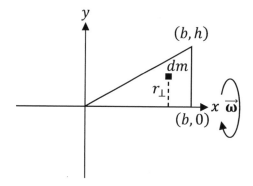

Perform the moment of inertia integral.

$$I = \int r_\perp^2 \, dm$$

For a solid triangle (as opposed to a wire bent into a triangle), we write $dm = \sigma dA$.

$$I = \int r_\perp^2 \sigma \, dA$$

Since the triangle is uniform, σ is a constant and may come out of the integral.

$$I = \sigma \int r_\perp^2 \, dA$$

For a triangle, we work with Cartesian coordinates and write the differential area element as $dA = dxdy$. This requires integrating over both x and y. If we let x vary from 0 to b, observe that for a given value of x, y varies from 0 to the hypotenuse of the triangle, which is less than h (except when x reaches its upper limit). See the diagram above. To find the proper upper limit for y, we need the equation of the line that serves as the triangle's hypotenuse. That line has a slope equal to $\frac{y_2 - y_1}{x_2 - x_1} = \frac{h - 0}{b - 0} = \frac{h}{b}$ and a y-intercept of 0. Since the general equation for a line is $y = mx + b$, the equation for this line is $y = \frac{hx}{b}$.

$$I = \sigma \int\limits_{x=0}^{b} \int\limits_{y=0}^{\frac{hx}{b}} r_{\perp}^2 \, dy \, dx$$

Don't pull $r_{\perp}^2$ out of the integral (since it isn't a constant). Note that $r_{\perp}$ extends from the axis of rotation to each dm. Study the previous diagram to see that $r_{\perp} = y$ since dm is a distance y from the axis of rotation (the x-axis). Substitute $r_{\perp}^2 = y^2$ into the integral.

$$I = \sigma \int\limits_{x=0}^{b} \int\limits_{y=0}^{\frac{hx}{b}} y^2 \, dy \, dx$$

First integrate over y since the upper limit of y $\left(\text{which equals } \frac{hx}{b}\right)$ is a variable.

$$I = \sigma \int\limits_{x=0}^{b} \left(\int\limits_{y=0}^{\frac{hx}{b}} y^2 \, dy \right) dx = \sigma \int\limits_{x=0}^{b} \left[\frac{y^3}{3}\right]_{y=0}^{y=\frac{hx}{b}} dx = \sigma \int\limits_{x=0}^{b} \left[\frac{\left(\frac{hx}{b}\right)^3}{3} - \frac{0^3}{3}\right] dx$$

$$I = \sigma \int\limits_{x=0}^{b} \frac{h^3 x^3}{3b^3} \, dx = \frac{\sigma h^3}{3b^3} \int\limits_{x=0}^{b} x^3 \, dx = \frac{\sigma h^3}{3b^3} \left[\frac{x^4}{4}\right]_{x=0}^{b} = \frac{\sigma h^3}{6b^3}\left(\frac{b^4}{4} - \frac{0^4}{4}\right) = \frac{\sigma h^3 b}{12}$$

In order to eliminate the constant σ from our answer, we will integrate over dm using the same substitutions as before, beginning with $dm = \sigma dA$.

$$m = \int dm = \int \sigma \, dA = \sigma \int dA = \sigma \int\limits_{x=0}^{b} \int\limits_{y=0}^{\frac{hx}{b}} dy \, dx$$

$$m = \sigma \int\limits_{x=0}^{b} \left(\int\limits_{y=0}^{\frac{hx}{b}} dy \right) dx = \sigma \int\limits_{x=0}^{b} [y]_{y=0}^{y=\frac{hx}{b}} dx = \sigma \int\limits_{x=0}^{b} \left(\frac{hx}{b} - 0\right) dx$$

$$m = \sigma \int\limits_{x=0}^{b} \frac{hx}{b} \, dx = \frac{\sigma h}{b} \int\limits_{x=0}^{b} x \, dx = \frac{\sigma h}{b} \left[\frac{x^2}{2}\right]_{x=0}^{b} = \frac{\sigma h}{b}\left(\frac{b^2}{2} - \frac{0^2}{2}\right) = \frac{\sigma h b}{2}$$

Solve for σ. Multiply both sides of the equation by 2 and divide hb.

$$\sigma = \frac{2m}{hb}$$

Substitute this expression for σ into the previous expression for moment of inertia.

$$I = \frac{\sigma h^3 b}{12} = \left(\frac{2m}{hb}\right)\frac{h^3 b}{12} = \frac{mh^2}{6}$$

The moment of inertia of the triangle about the x-axis is $I = \frac{mh^2}{6}$.

Example 196. Derive an equation for the moment of inertia of a solid uniform disc lying in the xy plane and centered about the origin about the y-axis in terms of its radius, R, and total mass, m.

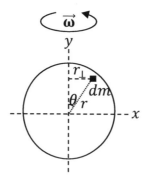

Solution. Begin with a labeled diagram. Draw a representative dm and label $r_\perp$, as we did in the diagram above. When we perform the integration, we effectively integrate over every dm that makes up the solid disc. Each dm has polar coordinates (r, θ). (For a review of polar coordinates, see Chapter 26.) Perform the moment of inertia integral.

$$I = \int r_\perp^2 \, dm$$

For a solid disc, we write $dm = \sigma dA$.

$$I = \int r_\perp^2 \sigma \, dA$$

Since the solid disc is uniform, σ is a constant and may come out of the integral.

$$I = \sigma \int r_\perp^2 \, dA$$

For a solid disc, the math is simpler if we work with 2D polar coordinates. The differential area element is $dA = rdrd\theta$. The limits of integration are from $r = 0$ to $r = R$ and $\theta = 0$ to $\theta = 2\pi$ **radians**. (The angular integration requires using radians, not degrees.)

$$I = \sigma \int_{r=0}^{R} \int_{\theta=0}^{2\pi} r_\perp^2 \, (rdr \, d\theta)$$

It's important to distinguish between $r, r_\perp$, and R.
- r extends from the origin to each dm. It is one of the two integration variables.
- $r_\perp$ extends from the axis of rotation (the y-axis). It is a variable, but different from r.
- R is the radius of the disc. It is a constant.

Don't pull $r_\perp^2$ out of the integral (since it isn't a constant). Note that $r_\perp$ extends from the axis of rotation to each dm. Study the previous diagram to see that $r_\perp = x$ since dm is a distance x from the axis of rotation (the y-axis). Substitute $r_\perp^2 = x^2$ into the integral.

$$I = \sigma \int_{r=0}^{R} \int_{\theta=0}^{2\pi} x^2 r \, dr \, d\theta$$

Use the equation $x = r \cos \theta$ for 2D polar coordinates (from Chapter 26).

$$I = \sigma \int_{r=0}^{R} \int_{\theta=0}^{2\pi} (r \cos\theta)^2 r \, dr \, d\theta = \sigma \int_{r=0}^{R} \int_{\theta=0}^{2\pi} r^3 \cos^2\theta \, dr \, d\theta$$

Note that one r came from $dA = r \, dr \, d\theta$, while the r^2 came from $x^2 = (r\cos\theta)^2 = r^2 \cos^2\theta$. Together, they make $r^2 r = r^3$. Also note that the cosine is squared because $r_\perp^2 = x^2 = (r\cos\theta)^2 = r^2\cos^2\theta$. You may do these integrals in any order since the limits are constant. We will integrate over θ first. Factor r out of the θ-integration (but not the r-integration).

$$I = \sigma \int_{r=0}^{R} r^3 \left(\int_{\theta=0}^{2\pi} \cos^2\theta \, d\theta \right) dr$$

The "trick" to this integral is to know the handy trig identity $\cos^2\theta = \frac{1+\cos 2\theta}{2}$.

$$I = \sigma \int_{r=0}^{R} r^3 \left(\int_{\theta=0}^{2\pi} \frac{1 + \cos 2\theta}{2} \, d\theta \right) dr = \sigma \int_{r=0}^{R} r^3 \left[\int_{\theta=0}^{2\pi} \left(\frac{1}{2} + \frac{\cos 2\theta}{2} \right) d\theta \right] dr$$

We will split the angular integral into two separate integrals using the rule $\int (y_1 + y_2) \, dx = \int y_1 \, dx + \int y_2 \, dx$, which in this case is $\int_{\theta=0}^{2\pi} \left(\frac{1}{2} + \frac{\cos 2\theta}{2} \right) d\theta = \int_{\theta=0}^{2\pi} \frac{1}{2} d\theta + \int_{\theta=0}^{2\pi} \frac{\cos 2\theta}{2} d\theta$.

$$I = \sigma \int_{r=0}^{R} r^3 \left(\int_{\theta=0}^{2\pi} \frac{1}{2} d\theta + \int_{\theta=0}^{2\pi} \frac{\cos 2\theta}{2} d\theta \right) dr = \sigma \int_{r=0}^{R} r^3 \left(\left[\frac{\theta}{2} \right]_{\theta=0}^{\theta=2\pi} + \left[\frac{\sin 2\theta}{4} \right]_{\theta=0}^{\theta=2\pi} \right) dr$$

Note that $\frac{d}{d\theta} \left(\frac{\sin 2\theta}{4} \right) = \frac{\cos 2\theta}{2}$ (the 2 comes out when you take a derivative of the argument, 2θ, after taking a derivative of the sine function) such that $\int \frac{\cos 2\theta}{2} d\theta = \frac{\sin 2\theta}{4}$.

$$I = \sigma \int_{r=0}^{R} r^3 \left[\frac{2\pi}{2} - \frac{0}{2} + \frac{\sin[2(2\pi)]}{4} - \frac{\sin[2(0)]}{4} \right] dr = \pi\sigma \int_{r=0}^{R} r^3 \, dr$$

Note that $\sin(4\pi) = \sin(720°) = 0$ and $\sin(0) = 0$.

$$I = \pi\sigma \left[\frac{r^4}{4} \right]_{r=0}^{r=R} = \pi\sigma \left(\frac{R^4}{4} - \frac{0^4}{4} \right) = \frac{\pi\sigma R^4}{4}$$

In order to eliminate the constant σ from our answer, we will integrate over dm using the same substitutions as before, beginning with $dm = \sigma dA$.

$$m = \int dm = \int \sigma \, dA = \sigma \int dA = \sigma \int_{r=0}^{R} \int_{\theta=0}^{2\pi} r \, dr \, d\theta = \sigma \int_{r=0}^{R} r \left(\int_{\theta=0}^{2\pi} d\theta \right) dr$$

$$m = \sigma \int_{r=0}^{R} r [\theta]_{\theta=0}^{\theta=2\pi} dr = \sigma \int_{r=0}^{R} 2\pi r \, dr = 2\pi\sigma \int_{r=0}^{R} r \, dr = 2\pi\sigma \left[\frac{r^2}{2} \right]_{r=0}^{r=R} = 2\pi\sigma \left(\frac{R^2}{2} - \frac{0^2}{2} \right)$$

$$m = \pi\sigma R^2$$

Solve for σ. Divide both sides of the equation by πR^2.

$$\sigma = \frac{m}{\pi R^2}$$

Substitute this expression for σ into the previous expression for moment of inertia.

$$I = \frac{\pi \sigma R^4}{4} = \frac{\pi R^4}{4}\left(\frac{m}{\pi R^2}\right) = \frac{mR^2}{4}$$

The moment of inertia of the solid disc about the y-axis is $I = \frac{mR^2}{4}$. Note that this is flipping mode, where $I = \frac{mR^2}{4}$, **not** to be confused with rolling mode, for which $I = \frac{mR^2}{2}$. (In rolling mode, the axis of rotation would be the z-axis, not the y-axis.)

Example 197. Derive an equation for the moment of inertia of a solid sphere with non-uniform density $\rho = \beta r$, where β is a positive constant, about an axis that passes through its center in terms of its radius, R, and total mass, m.

Solution. Begin with a labeled diagram. Draw a representative dm and label $r_\perp$, as we did in the diagram above. When we perform the integration, we effectively integrate over every dm that makes up the solid sphere. Each dm has spherical coordinates (r, θ, φ). (For a review of spherical coordinates, see Chapter 26.) This solution is simplest if the sphere is centered about the origin and if the axis of rotation is the z-axis.

$$I = \int r_\perp^2 \, dm$$

For a solid sphere (mathematicians call it a "ball"), we write $dm = \rho dV$.

$$I = \int r_\perp^2 \rho \, dV$$

Don't pull ρ or $r_\perp^2$ out of the integral. Substitute $\rho = \beta r$ (this was given in the problem).

$$I = \int r_\perp^2 \beta r \, dV$$

For a solid sphere, the math is simpler if we work with spherical coordinates. The differential volume element is $dV = r^2 \sin\theta \, dr d\theta d\varphi$. The limits of integration are from $r = 0$ to $r = R$, $\theta = 0$ to $\theta = \pi$ **radians**, and $\varphi = 0$ to $\varphi = 2\pi$ **radians**. (The angular integration requires using radians, not degrees.) Why does θ vary from 0 to π, whereas φ varies from 0 to 2π? Visualize the following. Integrating over $0 \le r \le R$ sweeps out a line from the origin. Integrating over $0 \le \varphi \le 2\pi$ sweeps the line (from the r-integration) into a cone.

It's then only necessary to integrate over $0 \leq \theta \leq \pi$ (and **not** up to 2π) to sweep the cone into a full sphere. One way to see this is to realize that the cone sweeps out a hemisphere once θ reaches $\frac{\pi}{2}$ radians (which is 90°).

$$I = \int_{r=0}^{R} \int_{\varphi=0}^{2\pi} \int_{\theta=0}^{\pi} r_\perp^2 (\beta r) \, (r^2 \sin\theta \, dr d\theta d\varphi)$$

Pull the constant β out of the integral (but don't pull anything else out of the integral because everything else is a variable). Note that $rr^2 = r^3$.

$$I = \beta \int_{r=0}^{R} \int_{\varphi=0}^{2\pi} \int_{\theta=0}^{\pi} r_\perp^2 \, r^3 \sin\theta \, dr d\theta d\varphi$$

It's important to distinguish between $r, r_\perp$, and R.

- r extends from the origin to each dm. It is one of the two integration variables.
- $r_\perp$ extends from the axis of rotation (the z-axis). It is a variable, but different from r.
- R is the radius of the sphere. It is a constant.

Don't pull $r_\perp^2$ out of the integral (since it isn't a constant). Note that $r_\perp$ extends from the axis of rotation to each dm. Study the previous diagram to see that $r_\perp = r \sin\theta$ since dm is a distance $r \sin\theta$ from the axis of rotation (the z-axis): $r_\perp$ is opposite to θ and the hypotenuse is the spherical coordinate r. Substitute $r_\perp^2 = r^2 \sin^2\theta$ into the integral.

$$I = \beta \int_{r=0}^{R} \int_{\varphi=0}^{2\pi} \int_{\theta=0}^{\pi} (r^2 \sin^2\theta) \, r^3 \sin\theta \, dr d\theta d\varphi = \beta \int_{r=0}^{R} \int_{\varphi=0}^{2\pi} \int_{\theta=0}^{\pi} r^5 \sin^3\theta \, dr d\theta d\varphi$$

Note that one r^2 came from $dV = r^2 \sin\theta \, dr d\theta d\varphi$, another r^2 came from $r_\perp^2 = r^2 \sin^2\theta$, and an r came from $\rho = \beta r$. Together, they make $r^2 r^2 r = r^5$. Also note that one $\sin\theta$ came from $dV = r^2 \sin\theta \, dr d\theta d\varphi$ and a $\sin^2\theta$ came from $r_\perp^2 = r^2 \sin^2\theta$. Together, they make $\sin\theta \sin^2\theta = \sin^3\theta$. We can pull r out of the angular integrals (but not the r-integral), and we can pull $\sin\theta$ out of the r and φ integrals (but not the θ-integral).

$$I = \beta \int_{r=0}^{R} r^5 dr \int_{\theta=0}^{\pi} \sin^3\theta \, d\theta \int_{\varphi=0}^{2\pi} d\varphi = \beta \left[\frac{r^6}{6} \right]_{r=0}^{R} [\varphi]_{\varphi=0}^{2\pi} \int_{\theta=0}^{\pi} \sin^3\theta \, d\theta$$

$$I = \frac{2\pi\beta R^6}{6} \int_{\theta=0}^{\pi} \sin^3\theta \, d\theta = \frac{\pi\beta R^6}{3} \int_{\theta=0}^{\pi} \sin^3\theta \, d\theta$$

The 'trick' to performing the θ-integration is to use the following substitution:

$$\sin^3\theta = \sin\theta \sin^2\theta = \sin\theta \,(1 - \cos^2\theta) = \sin\theta - \sin\theta \cos^2\theta$$

(Here, we used the fact that $\sin^2\theta + \cos^2\theta = 1$ to write $\sin^2\theta = 1 - \cos^2\theta$.) After we plug this expression into the previous integral, we will separate the integral into two terms.

$$I = \frac{\pi\beta R^6}{3} \int_{\theta=0}^{\pi} \sin^3\theta \, d\theta = \frac{\pi\beta R^6}{3} \left(\int_{\theta=0}^{\pi} \sin\theta \, d\theta - \int_{\theta=0}^{\pi} \sin\theta \cos^2\theta \, d\theta \right)$$

267

Recall from Chapter 26 that an integral can often be made simpler with a substitution. With the second integral, we define $u = \cos\theta$ such that $du = -\sin\theta\,d\theta$. When $\theta = 0$, the new variable u equals $\cos 0 = 1$, and when $\theta = \pi$, the variable u equals $\cos\pi = -1$.

$$I = \frac{\pi\beta R^6}{3}\left([-\cos\theta]_{\theta=0}^{\pi} + \int_{u=1}^{-1} u^2\,du\right) = \frac{\pi\beta R^6}{3}\left(-\cos\pi + \cos 0 + \left[\frac{u^3}{3}\right]_{u=1}^{-1}\right)$$

$$I = \frac{\pi\beta R^6}{3}\left[-(-1) + 1 + \frac{(-1)^3}{3} - \frac{(1)^3}{3}\right] = \frac{\pi\beta R^6}{3}\left(1 + 1 - \frac{1}{3} - \frac{1}{3}\right) = \frac{\pi\beta R^6}{3}\frac{4}{3} = \frac{4\pi\beta R^6}{9}$$

In order to eliminate the constant β from our answer, we will integrate over dm using the same substitutions as before, beginning with $dm = \rho\,dV$.

$$m = \int dm = \int \rho\,dV = \int_{r=0}^{R}\int_{\varphi=0}^{2\pi}\int_{\theta=0}^{\pi}(\beta r)\,(r^2\sin\theta\,dr d\theta d\varphi)$$

$$m = \beta\int_{r=0}^{R}\int_{\varphi=0}^{2\pi}\int_{\theta=0}^{\pi} r^3\sin\theta\,dr d\varphi d\theta = \beta\int_{r=0}^{R}r^3\,dr\int_{\theta=0}^{\pi}\sin\theta\,d\theta\int_{\varphi=0}^{2\pi}d\varphi$$

$$m = \beta\left[\frac{r^4}{4}\right]_{r=0}^{R}[-\cos\theta]_{\theta=0}^{\pi}[\varphi]_{\varphi=0}^{2\pi} = \frac{\beta R^4}{4}[-\cos\pi - (-\cos 0)](2\pi) = \frac{\pi\beta R^4}{2}(1+1)$$

$$m = \pi\beta R^4$$

Solve for β. Divide both sides of the equation by πR^4.

$$\beta = \frac{m}{\pi R^4}$$

Substitute this expression for β into the previous expression for moment of inertia.

$$I = \frac{4\pi\beta R^6}{9} = \frac{4\pi R^6}{9}\left(\frac{m}{\pi R^4}\right) = \frac{4mR^2}{9}$$

The moment of inertia of the solid sphere about the z-axis is $I = \frac{4mR^2}{9}$. Recall that this sphere was **non-uniform**. (If it were uniform, the answer would be $I = \frac{2mR^2}{5}$.)

Example 198. Derive an equation for the moment of inertia of a very thin uniform hollow sphere about an axis that passes through its center in terms of the radius, R, and total mass, m, of the hollow sphere.

Solution. The key to this problem is to understand how it is different from the previous example. It's instructive to compare the wording for these two problems. First of all, for a very thin hollow sphere, $r = R$ is a constant (whereas r was a variable for the solid sphere in the previous example), so we will have a double integral over the surface area ($dA = R^2\sin\theta\,d\theta d\varphi$) instead of a triple integral over the volume ($dV = r^2\sin\theta\,dr d\theta d\varphi$). Another important difference is that the hollow sphere is uniform in this example, so that we may pull the 2D density σ out of the integral (whereas the 3D density ρ in the previous example was non-uniform and thus couldn't be pulled out of the integral).

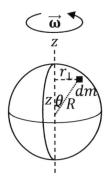

Begin with a labeled diagram. Draw a representative dm and label $r_\perp$, as we did in the diagram above. When we perform the integration, we effectively integrate over every dm that makes up the sphere. Each dm has spherical coordinates (r, θ, φ). Note that $r = R$ in this example because every dm lies on the surface of the hollow sphere. The solution is simplest if the sphere is centered about the origin and if the axis of rotation is the z-axis.

$$I = \int r_\perp^2 \, dm$$

For a very thin hollow sphere, we write $dm = \sigma dA$.

$$I = \int r_\perp^2 \sigma \, dA$$

Since the hollow sphere is uniform, σ is a constant and may come out of the integral.

$$I = \sigma \int r_\perp^2 \, dA$$

For a hollow sphere, the math is simpler if we work with spherical coordinates. The differential area element is $dA = R^2 \sin\theta \, d\theta d\varphi$. The limits of integration are from $\theta = 0$ to $\theta = \pi$ **radians** and $\varphi = 0$ to $\varphi = 2\pi$ **radians**. (These are the same limits as in the previous example, except that $r = R$ is constant for the very thin hollow sphere.)

$$I = \sigma \int_{\varphi=0}^{2\pi} \int_{\theta=0}^{\pi} r_\perp^2 \, (R^2 \sin\theta \, d\theta d\varphi)$$

Don't pull $r_\perp^2$ out of the integral (since it isn't a constant). Note that $r_\perp$ extends from the axis of rotation to each dm. Study the previous diagram to see that $r_\perp = R\sin\theta$ since dm is a distance $R\sin\theta$ from the axis of rotation (the z-axis): $r_\perp$ is opposite to θ and the hypotenuse is the radius R. Substitute $r_\perp^2 = R^2 \sin^2\theta$ into the integral.

$$I = \sigma \int_{\varphi=0}^{2\pi} \int_{\theta=0}^{\pi} (R^2 \sin^2\theta) \, (R^2 \sin\theta \, d\theta d\varphi) = \sigma \int_{\varphi=0}^{2\pi} \int_{\theta=0}^{\pi} R^4 \sin^3\theta \, d\theta d\varphi$$

Pull the constant R out of the integration. We can pull $\sin\theta$ out of the φ integral (but not the θ-integral).

$$I = \sigma R^4 \int_{\theta=0}^{\pi} \sin^3\theta \left(\int_{\varphi=0}^{2\pi} d\varphi \right) d\theta = 2\pi\sigma R^4 \int_{\theta=0}^{\pi} \sin^3\theta \, d\theta$$

The 'trick' to performing the θ-integration is to use the following substitution:
$$\sin^3 \theta = \sin \theta \sin^2 \theta = \sin \theta \,(1 - \cos^2 \theta) = \sin \theta - \sin \theta \cos^2 \theta$$
(Here, we used the fact that $\sin^2 \theta + \cos^2 \theta = 1$ to write $\sin^2 \theta = 1 - \cos^2 \theta$.) After we plug this expression into the previous integral, we will separate the integral into two terms.
$$I = 2\pi\sigma R^4 \int_{\theta=0}^{\pi} \sin^3 \theta \, d\theta = 2\pi\sigma R^4 \left(\int_{\theta=0}^{\pi} \sin \theta \, d\theta - \int_{\theta=0}^{\pi} \sin \theta \cos^2 \theta \, d\theta \right)$$
Recall from Chapter 26 that an integral can often be made simpler with a substitution. With the second integral, we define $u = \cos \theta$ such that $du = -\sin \theta \, d\theta$. When $\theta = 0$, the new variable u equals $\cos 0 = 1$, and when $\theta = \pi$, the variable u equals $\cos \pi = -1$.
$$I = 2\pi\sigma R^4 \left([-\cos \theta]_{\theta=0}^{\pi} + \int_{u=1}^{-1} u^2 du \right) = 2\pi\sigma R^4 \left(-\cos \pi + \cos 0 + \left[\frac{u^3}{3}\right]_{u=1}^{-1} \right)$$
$$I = 2\pi\sigma R^4 \left[-(-1) + 1 + \frac{(-1)^3}{3} - \frac{(1)^3}{3} \right] = 2\pi\sigma R^4 \left(1 + 1 - \frac{1}{3} - \frac{1}{3} \right) = 2\pi\sigma R^4 \frac{4}{3} = \frac{8\pi\sigma R^4}{3}$$
In order to eliminate the constant σ from our answer, we will integrate over dm using the same substitutions as before, beginning with $dm = \sigma dA$.
$$m = \int dm = \int \sigma \, dA = \sigma \int dA = \sigma \int_{\varphi=0}^{2\pi} \int_{\theta=0}^{\pi} R^2 \sin \theta \, d\theta \, d\varphi = \sigma R^2 \int_{\theta=0}^{\pi} \sin \theta \, d\theta \int_{\varphi=0}^{2\pi} d\varphi$$
$$m = \sigma R^2 [-\cos \theta]_{\theta=0}^{\pi} [\varphi]_{\varphi=0}^{2\pi} = \sigma R^2 [-\cos \pi - (-\cos 0)](2\pi) = 2\pi\sigma R^2 (1 + 1)$$
$$m = 4\pi\sigma R^2$$
Solve for σ. Divide both sides of the equation by $4\pi R^4$.
$$\sigma = \frac{m}{4\pi R^4}$$
Substitute this expression for σ into the previous expression for moment of inertia.
$$I = \frac{8\pi\sigma R^4}{3} = \frac{8\pi R^4}{3} \left(\frac{m}{4\pi R^4}\right) = \frac{2mR^2}{3}$$
The moment of inertia of the very thin hollow sphere about the z-axis is $I = \frac{2mR^2}{3}$. (Note that this sphere is hollow and very thin. If it were a solid sphere instead, the answer would be $I = \frac{2mR^2}{5}$.)

33 A PULLEY ROTATING WITHOUT SLIPPING

Newton's Second Law and the Sum of the Torques
$$\sum F_x = ma_x \quad , \quad \sum F_y = ma_y \quad , \quad \sum \tau = I\alpha$$
Torque
$$\tau = r\,F\sin\theta$$
Angular Acceleration
$$\alpha = \frac{a_x}{R_p}$$

Symbol	Name	Units
m	mass	kg
a	acceleration	m/s^2
a_x	x-component of acceleration	m/s^2
a_y	y-component of acceleration	m/s^2
F	force	N
I	moment of inertia	kg·m^2
α	angular acceleration	rad/s^2
τ	torque	N·m
r	distance from the axis of rotation to $\vec{F}$	m
θ	the angle between $\vec{r}$ and $\vec{F}$	° or rad
R_p	radius of the pulley	m

Example 199. As illustrated below, a 50-kg monkey is connected to a 40-kg box of bananas by a cord that passes over a pulley. The 20-kg pulley is a solid disc. The cord rotates with the pulley without slipping. The coefficient of friction between the box and ground is $\frac{1}{2}$. Determine the acceleration of the system. The system begins from rest.

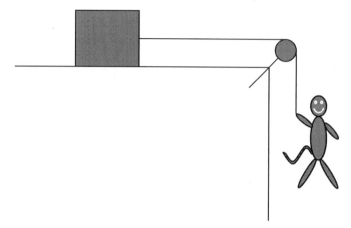

Solution. Draw and label a free-body diagram (FBD) for the box and monkey and also draw and label an extended FBD for the pulley showing where each force acts on the pulley.

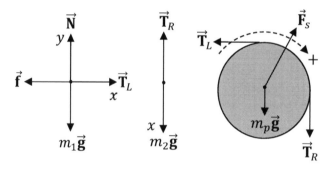

- Each object has **weight** ($m_1\vec{g}$ and $m_2\vec{g}$) pulling straight down. The weight of the pulley ($m_p\vec{g}$) acts on the center of the pulley (where gravity acts on "average").
- There are two pairs of **tension** ($\vec{T}_L$ and $\vec{T}_R$) forces: one tension in each section of the cord. Friction between the cord and the pulley allows the cord to rotate with the pulley without slipping (as stated in the problem). The two tensions acting on the pulley must be different in order for the pulley to develop the angular acceleration needed to match the cord's acceleration. Note that Newton's third law still applies as there are two equal and opposite tension force pairs.
- A **normal force** ($\vec{N}$) supports the box (perpendicular to the surface). Since the surface is horizontal, normal force points up.
- **Friction** ($\vec{f}$) acts opposite to the velocity of the box of bananas: to the left. (Although the box begins from rest, it will travel to the right once the motion starts.)
- A support force ($\vec{F}_s$) holds the pulley in place. It must be up and to the right in order to cancel the other forces acting on the pulley.

- Label $+x$ in the direction that each object accelerates: to the right for the box and down for the monkey. Think of the pulley as bending the x-axis.
- Label $+y$ perpendicular to x. For the box of bananas, $+y$ is up.
- The positive sense of rotation for the pulley must match the choice of $+x$. Since the box will travel right and the monkey will fall downward, the pulley will rotate clockwise. We must label the positive sense of rotation for the pulley as clockwise.

The monkey and box have the same acceleration because they are connected by a cord (which we will assume doesn't stretch). Apply Newton's second law to each object. We don't need $\sum F_{2y}$ because there are no forces pulling on the monkey with a horizontal component.

$$\sum F_{1x} = m_1 a_x \quad , \quad \sum F_{1y} = m_1 a_y \quad , \quad \sum F_{2x} = m_2 a_x$$

Rewrite each sum using the forces from the FBD's. Note that $a_y = 0$ because the box doesn't accelerate vertically: The box has a_x, but not a_y.

$$T_L - f = m_1 a_x \quad , \quad N - m_1 g = 0 \quad , \quad m_2 g - T_R = m_2 a_x$$

We must also sum the torques acting on the pulley.

$$\sum \tau = I \alpha$$

The pulley's weight and the support force don't exert torques (they don't affect the pulley's rotation). The right tension creates a positive torque (clockwise) while the left tension creates a negative torque (counterclockwise). For the tensions, $\theta = 90°$ in the torque equation ($\tau = r F \sin \theta$) since tension pulls tangentially to the pulley. Since the pulley is a solid disc, its moment of inertia is $I = \frac{1}{2} m_p R_p^2$ (Chapter 32).

$$R_p T_R \sin 90° - R_p T_L \sin 90° = \frac{1}{2} m_p R_p^2 \alpha$$

Divide both sides by R_p. Note that $\sin 90° = 1$.

$$T_R - T_L = \frac{1}{2} m_p R_p \alpha$$

Make the substitution $R_p \alpha = a_x$ since the angular acceleration of the pulley must match the acceleration of the masses and the pulley.

$$T_R - T_L = \frac{1}{2} m_p a_x$$

When there is friction in a problem, solve for normal force in the y-sum.

$$N - m_1 g = 0$$
$$N = m_1 g = 40(9.81) = 392 \text{ N}$$

Then use the equation for friction.

$$f = \mu N = \frac{1}{2}(392) = 196 \text{ N}$$

Add the two equations from the x-sums ($T_L - f = m_1 a_x$ and $m_2 g - T_R = m_2 a_x$) to the simplified equation from the sum of the torques ($T_R - T_L = \frac{1}{2} m_p a_x$) in order to cancel tension. The sum of the left-hand sides equals the sum of the right-hand sides.

$$T_L - f + m_2 g - T_R + T_R - T_L = m_1 a_x + m_2 a_x + \frac{1}{2} m_p a_x$$

$$-f + m_2 g = m_1 a_x + m_2 a_x + \frac{1}{2} m_p a_x$$

$$-196 + 50(9.81) = 40 a_x + 50 a_x + \frac{1}{2}(20) a_x$$

$$-196 + 491 = 40 a_x + 50 a_x + 10 a_x$$

$$295 = 100 a_x$$

$$a_x = \frac{295}{100} = 3.0 \text{ m/s}^2$$

The acceleration of the system is $a_x = 3.0$ m/s^2.

Example 200. As illustrated below, two monkeys are connected by a cord. The 20-kg pulley is a solid disc. The cord rotates with the pulley without slipping. The surface is frictionless. Determine the acceleration of the system. The system begins from rest.

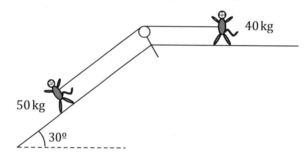

Solution. Draw and label a free-body diagram (FBD) for each monkey and also draw and label an extended FBD for the pulley showing where each force acts on the pulley.

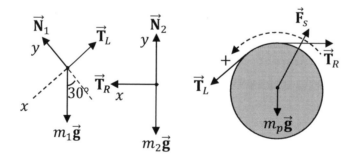

- Each object has **weight** ($m_1 \vec{\mathbf{g}}$, $m_2 \vec{\mathbf{g}}$, and $m_p \vec{\mathbf{g}}$) pulling straight down.

- As explained in the previous example, there are two pairs of **tension** ($\vec{\mathbf{T}}_L$ and $\vec{\mathbf{T}}_R$) forces: one tension in each section of the cord.

- **Normal forces** ($\vec{\mathbf{N}}_1$ and $\vec{\mathbf{N}}_2$) push perpendicular to the surfaces.

- A support force ($\vec{\mathbf{F}}_s$) holds the pulley in place. It must be up and to the right in order to cancel the other forces acting on the pulley.

- Label $+x$ in the direction that each object accelerates: down the incline for one and to the left for the other. Think of the pulley as bending the x-axis.
- Label $+y$ perpendicular to x: $+y$ is along the normal forces.
- The positive sense of rotation for the pulley must match the choice of $+x$. Since the system will travel left, the pulley will rotate counterclockwise. We must label the positive sense of rotation for the pulley as counterclockwise.

The monkeys have the same acceleration because they are connected by a cord (which we will assume doesn't stretch). Apply Newton's second law to each monkey.

$$\sum F_{1x} = m_1 a_x \quad , \quad \sum F_{1y} = m_1 a_y \quad , \quad \sum F_{2x} = m_2 a_x \quad , \quad \sum F_{2y} = m_2 a_y$$

Rewrite each sum using the forces from the FBD's. Note that $a_y = 0$ because the monkeys don't accelerate perpendicular to the surfaces: The monkeys have a_x, but not a_y. Example 94 in Chapter 14 explains which sums $m_1 g \sin 30°$ and $m_1 g \cos 30°$ go in.

$$m_1 g \sin 30° - T_L = m_1 a_x \quad , \quad N_1 - m_1 g \cos 30° = 0 \quad , \quad T_R = m_2 a_x \quad , \quad N_2 - m_{2g} = 0$$

We must also sum the torques acting on the pulley.

$$\sum \tau = I\alpha$$

The pulley's weight and the support force don't exert torques (they don't affect the pulley's rotation). The right tension creates a positive torque (counterclockwise) while the left tension creates a negative torque (clockwise). For the tensions, $\theta = 90°$ in the torque equation ($\tau = r F \sin \theta$) since tension pulls tangentially to the pulley. Since the pulley is a solid disc, its moment of inertia is $I = \frac{1}{2} m_p R_p^2$ (Chapter 32).

$$R_p T_L \sin 90° - R_p T_R \sin 90° = \frac{1}{2} m_p R_p^2 \alpha$$

Divide both sides by R_p. Note that $\sin 90° = 1$.

$$T_L - T_R = \frac{1}{2} m_p R_p \alpha$$

Make the substitution $R_p \alpha = a_x$ since the angular acceleration of the pulley must match the acceleration of the monkeys and the pulley.

$$T_L - T_R = \frac{1}{2} m_p a_x$$

Add the two equations from the x-sums ($m_1 g \sin 30° - T_L = m_1 a_x$ and $T_R = m_2 a_x$) to the simplified equation from the sum of the torques ($T_L - T_R = \frac{1}{2} m_p a_x$) in order to cancel tension. The sum of the left-hand sides equals the sum of the right-hand sides.

$$m_1 g \sin 30° - T_L + T_R + T_L - T_R = m_1 a_x + m_2 a_x + \frac{1}{2} m_p a_x$$

$$m_1 g \sin 30° = m_1 a_x + m_2 a_x + \frac{1}{2} m_p a_x$$

$$50(9.81)\left(\frac{1}{2}\right) = 50 a_x + 40 a_x + \frac{1}{2}(20) a_x$$

$$245 = 40a_x + 50a_x + 10a_x$$
$$245 = 100a_x$$
$$a_x = \frac{245}{100} = 2.5 \text{ m/s}^2$$

The acceleration of the system is $a_x = 2.5 \text{ m/s}^2 = \frac{5}{2} \text{ m/s}^2$. Note that if you solve for the tensions, you get $T_R = 100$ N and $T_L = 125$ N. Since the horizontal component of T_L, which equals $T_{Lx} = T_L \cos 30° = (125)\frac{\sqrt{3}}{2} = 108$ N, is greater than the horizontal component of T_R, which is $T_{Rx} = T_R = 100$ N since T_R is horizontal, this is how we know that the support force, F_s, is up to the right (and not up to the left or straight up).

34 ROLLING WITHOUT SLIPPING

Translational Kinetic Energy

$$KE_{t0} = \frac{1}{2}mv_0^2 \quad , \quad KE_t = \frac{1}{2}mv^2$$

Rotational Kinetic Energy

$$KE_{r0} = \frac{1}{2}I\omega_0^2 \quad , \quad KE_r = \frac{1}{2}I\omega^2$$

Total Kinetic Energy

$$KE_0 = KE_{t0} + KE_{r0} = \frac{1}{2}mv_0^2 + \frac{1}{2}I\omega_0^2 \quad , \quad KE = KE_t + KE_r = \frac{1}{2}mv^2 + \frac{1}{2}I\omega^2$$

Conservation of Energy

$$PE_0 + KE_0 + W_{nc} = PE + KE$$

Gravitational Potential Energy

$$PE_{g0} = mgh_0 \quad , \quad PE_g = mgh$$

Spring Potential Energy

$$PE_{s0} = \frac{1}{2}kx_0^2 \quad , \quad PE_s = \frac{1}{2}kx^2$$

Angular Speed

$$\omega_0 = \frac{v_0}{R} \quad , \quad \omega = \frac{v}{R}$$

Symbol	Name	SI Units
PE_0	initial potential energy	J
PE	final potential energy	J
KE_0	initial (total) kinetic energy	J
KE	final (total) kinetic energy	J
KE_{t0}	initial translational kinetic energy	J
KE_t	final translational kinetic energy	J
KE_{r0}	initial rotational kinetic energy	J
KE_r	final rotational kinetic energy	J
W_{nc}	nonconservative work	J
m	mass	kg
g	gravitational acceleration	m/s^2
h_0	initial height (relative to the reference height)	m
h	final height (relative to the reference height)	m
v_0	initial speed	m/s
v	final speed	m/s
I	moment of inertia	$kg \cdot m^2$
ω_0	initial angular speed	rad/s
ω	final angular speed	rad/s
k	spring constant	N/m or kg/s^2
x_0	initial displacement of a spring from equilibrium	m
x	final displacement of a spring from equilibrium	m

Example 201. A hollow sphere rolls without slipping down an incline from rest. The hollow sphere reaches the bottom of the incline after descending a height of 75 m. Determine the speed of the hollow sphere as it reaches the bottom of the incline.

Solution. Begin by labeling a diagram. The initial position (i) is where it starts, while the final position (f) is just before the sphere reaches the bottom of the incline (the final speed is **not** zero). We choose the reference height (RH) to be at the bottom of the incline.

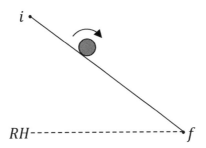

Consider the terms of the conservation of energy equation.

- There is initial gravitational potential energy ($PE_0 = mgh_0$) because the sphere is above the reference height (RH) at i.
- There is no initial kinetic energy ($KE_0 = 0$) because the sphere begins from rest.
- There must be friction in order for the sphere to roll without slipping, yet $W_{nc} = 0$ because in the case of **rolling** (unlike sliding), friction doesn't subtract mechanical energy from the system.
- There is no final potential energy ($PE = 0$) since f is at the same height as RH.
- There is final kinetic energy ($KE = \frac{1}{2}mv^2 + \frac{1}{2}I\omega^2$) since the sphere is **rolling** at f. (The sphere has $\frac{1}{2}mv^2$ because its center of mass is moving and it has $\frac{1}{2}I\omega^2$ because it is rotating.)

Write out conservation of energy for the sphere.

$$PE_0 + KE_0 + W_{nc} = PE + KE$$

Substitute the previous expressions into the conservation of energy equation.

$$mgh_0 + 0 + 0 = 0 + \frac{1}{2}mv^2 + \frac{1}{2}I\omega^2$$

Look up the equation for the moment of a inertia of a very thin hollow sphere (Chapter 32): $I = \frac{2}{3}mR^2$.

$$mgh_0 = \frac{1}{2}mv^2 + \frac{1}{2}\left(\frac{2}{3}mR^2\right)\omega^2$$

$$mgh_0 = \frac{1}{2}mv^2 + \frac{1}{3}mR^2\omega^2$$

Divide both sides of the equation by mass. The mass cancels out.

$$gh_0 = \frac{1}{2}v^2 + \frac{1}{3}R^2\omega^2$$

Use the equation $\omega = \frac{v}{R}$ (after which R^2 will cancel). Note that $\omega^2 = \frac{v^2}{R^2}$.

$$gh_0 = \frac{1}{2}v^2 + \frac{1}{3}R^2\left(\frac{v^2}{R^2}\right)$$

$$gh_0 = \frac{1}{2}v^2 + \frac{1}{3}v^2$$

Factor out the v^2. Add the fractions with a **common denominator**.

$$gh_0 = v^2\left(\frac{1}{2} + \frac{1}{3}\right)$$

$$gh_0 = v^2\left(\frac{3}{6} + \frac{2}{6}\right)$$

$$gh_0 = \frac{5}{6}v^2$$

Multiply both sides of the equation by 6 and divide by 5.

$$v^2 = \frac{6gh_0}{5}$$

Squareroot both sides of the equation.

$$v = \sqrt{\frac{6gh_0}{5}} = \sqrt{\frac{6(9.81)(75)}{5}} = \sqrt{883} = 30 \text{ m/s}$$

The speed of the hollow sphere when it reaches the bottom of the incline is $v = 30$ m/s.

Example 202. A donut for which $I = \frac{3mR^2}{4}$ rolls without slipping up a 30° incline with an initial speed of 40 m/s. How far does the donut travel up the incline?

Solution. Begin by labeling a diagram. The initial position (i) is when the donut begins rolling at the bottom, while the final position (f) is when it reaches its highest position. We choose the reference height (RH) to be at the bottom of the incline.

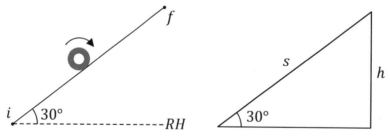

Consider the terms of the conservation of energy equation.

- There is no initial potential energy ($PE_0 = 0$) since i is at the same height as RH.
- There is initial kinetic energy ($KE_0 = \frac{1}{2}mv_0^2 + \frac{1}{2}I\omega_0^2$) since the donut is **rolling** at i. (The donut has $\frac{1}{2}mv_0^2$ because its center of mass is moving and it has $\frac{1}{2}I\omega_0^2$ because it is rotating.).
- There must be friction in order for the donut to roll without slipping, yet $W_{nc} = 0$ because in the case of **rolling** (unlike sliding), friction doesn't subtract mechanical energy from the system.

- There is final gravitational potential energy ($PE = mgh$) because the donut is above the reference height (RH) at f.
- There is no final kinetic energy ($KE = 0$) because the donut runs out of speed at f (otherwise, the donut would continue to rise higher). Note that the final position (f) is where the box reaches its highest point.

Write out conservation of energy for the donut.

$$PE_0 + KE_0 + W_{nc} = PE + KE$$

Substitute the previous expressions into the conservation of energy equation.

$$0 + \frac{1}{2}mv_0^2 + \frac{1}{2}I\omega_0^2 + 0 = mgh + 0$$

Plug in the expression for the moment of inertia given in the problem: $I = \frac{3mR^2}{4}$. (This equation isn't true for all donuts, but applies to this particular donut.)

$$\frac{1}{2}mv_0^2 + \frac{1}{2}\left(\frac{3mR^2}{4}\right)\omega_0^2 = mgh$$

$$\frac{1}{2}mv_0^2 + \frac{3mR^2}{8}\omega_0^2 = mgh$$

Divide both sides of the equation by mass. The mass cancels out.

$$\frac{1}{2}v_0^2 + \frac{3R^2}{8}\omega_0^2 = gh$$

Use the equation $\omega_0 = \frac{v_0}{R}$ (after which R^2 will cancel). Note that $\omega_0^2 = \frac{v_0^2}{R^2}$.

$$\frac{1}{2}v_0^2 + \frac{3R^2}{8}\left(\frac{v_0^2}{R^2}\right) = gh$$

$$\frac{1}{2}v_0^2 + \frac{3}{8}v_0^2 = gh$$

Factor out the v_0^2. Add the fractions with a **common denominator**.

$$\left(\frac{1}{2} + \frac{3}{8}\right)v_0^2 = gh$$

$$\left(\frac{4}{8} + \frac{3}{8}\right)v_0^2 = gh$$

$$\frac{7}{8}v_0^2 = gh$$

Divide both sides of the equation by gravitational acceleration.

$$h = \frac{7v_0^2}{8g}$$

Note that the problem asked for the distance (s) that the donut travels up the incline, not the final height (h) of the donut. Study the right triangle that appears near the beginning of this solution: The side opposite to $30°$ is h, while s is the hypotenuse of the triangle. Apply the sine function from trigonometry to relate h to s.

$$\sin 30° = \frac{h}{s}$$

Multiply both sides of the equation by s.

$$h = s \sin 30° = s\left(\frac{1}{2}\right) = \frac{s}{2}$$

Substitute this equation for height into the previous equation for height that we obtained from conservation of energy.

$$\frac{s}{2} = \frac{7v_0^2}{8g}$$

Multiply both sides of the equation by 2.

$$s = \frac{7v_0^2}{4g} = \frac{7(40)^2}{4(9.81)} = 285 \text{ m}$$

The distance the donut travels up the incline is $s = 285$ m (before it rolls back down).

35 CONSERVATION OF ANGULAR MOMENTUM

Angular Momentum
$\vec{L} = I\vec{\omega}$
Conservation of Angular Momentum
$I_{10}\vec{\omega}_{10} + I_{20}\vec{\omega}_{20} = I_1\vec{\omega}_1 + I_2\vec{\omega}_2$

Symbol	Name	SI Units
$\vec{L}$	angular momentum	$\text{kg·m}^2/\text{s}$
I	moment of inertia	kg·m^2
$\vec{\omega}$	angular velocity	rad/s
ω	angular speed	rad/s
I_{10}	initial moment of inertia of object 1	kg·m^2
I_{20}	initial moment of inertia of object 2	kg·m^2
I_1	final moment of inertia of object 1	kg·m^2
I_2	final moment of inertia of object 2	kg·m^2
$\vec{\omega}_{10}$	initial angular velocity of object 1	rad/s
$\vec{\omega}_{20}$	initial angular velocity of object 2	rad/s
$\vec{\omega}_1$	final angular velocity of object 1	rad/s
$\vec{\omega}_2$	final angular velocity of object 2	rad/s
m_1	mass of object 1	kg
m_2	mass of object 2	kg
R_1	radius of object 1	m
R_2	radius of object 2	m

Example 203. A 30-kg monkey is placed at rest at the center of a merry-go-round. The

merry-go-round is a large solid disc which has a mass of 120 kg and a diameter of 16 m. A gorilla spins the merry-go-round at a rate of $\frac{1}{4}$ rev/s and lets go. As the merry-go-round spins, the monkey walks outward until he reaches the edge. Find the angular speed of the merry-go-round when the monkey reaches the edge. Neglect any friction with the axle.

Solution. Since the net torque acting on the system (the monkey plus the merry-go-round) equals zero (after the gorilla lets go), we may apply the law of conservation of angular momentum.

$$I_{10}\vec{\omega}_{10} + I_{20}\vec{\omega}_{20} = I_1\vec{\omega}_1 + I_2\vec{\omega}_2$$

From Chapter 32, the moment of inertia of the solid disc is $\frac{1}{2}m_1 R_1^2$ and the moment of inertia of the pointlike monkey is $m_2 R_2^2$. Initially, $R_{20} = 0$ because the monkey is at the center, while finally $R_2 = \frac{D}{2} = \frac{16}{2} = 8.0$ m when the monkey reaches the edge. Note that $R_1 = \frac{D}{2} = \frac{16}{2} = 8.0$ m for the disc. Plug this information into the equation for conservation of angular momentum.

$$\frac{1}{2}m_1 R_1^2 \vec{\omega}_{10} + 0 = \frac{1}{2}m_1 R_1^2 \vec{\omega}_1 + m_2 R_2^2 \vec{\omega}_2$$

$$\frac{1}{2}(120)(8)^2 \vec{\omega}_{10} + 0 = \frac{1}{2}(120)(8)^2 \vec{\omega}_1 + (30)(8)^2 \vec{\omega}_2$$

$$(60)(64)\,\vec{\omega}_{10} + 0 = (60)(64)\,\vec{\omega}_1 + (30)(64)\,\vec{\omega}_2$$

$$3840\,\vec{\omega}_{10} + 0 = 3840\,\vec{\omega}_1 + 1920\,\vec{\omega}_2$$

The initial angular speed of the merry-go-round is $\omega_{10} = \frac{1}{4}$ rev/s. We don't need to convert to radians for this equation so long as we're consistent (don't mix and match revolutions with radians). The merry-go-round and monkey have the same final angular speed ($\omega_1 = \omega_2$), so we'll just call them both ω.

$$3840\left(\frac{1}{4}\right) = 3840\omega + 1920\omega$$

Factor out the final angular speed.

$$960 = (3840 + 1920)\omega$$

$$960 = 5760\omega$$

Divide both sides of the equation by 5760.

$$\omega = \frac{960}{5760} = \frac{960 \div 160}{5760 \div 160} = \frac{1}{6}\ \text{rev/s}$$

The final angular speed is $\omega = \frac{1}{6}$ rev/s. The answer came out in rev/s because we put the initial angular speed in rev/s.

Example 204. A monkey is SO frustrated with his slow internet connection that he picks up his laptop, slams it against a brick wall, and then jumps high into the air and stomps on it. At that exact moment, the earth suddenly contracts until it has one-third of its initial radius. How long will a 'day' be now?

Solution. Since the net torque acting on the system (the earth) equals zero (since the average force of contraction is directed toward the center of the earth), we may apply the law of conservation of angular momentum. The earth is the only object in the system.

$$I_0 \vec{\omega}_0 = I \vec{\omega}$$

Treat the earth as roughly a uniform solid sphere (Chapter 32): $I_0 = \frac{2}{5} m R_0^2$ and $I = \frac{2}{5} m R^2$.

$$\frac{2}{5} m R_0^2 \omega_0 = \frac{2}{5} m R^2 \omega$$

Divide both sides of the equation by $\frac{2m}{5}$. Mass and the coefficient cancel out.

$$R_0^2 \omega_0 = R^2 \omega$$

Divide both sides of the equation by R^2.

$$\omega = \left(\frac{R_0}{R}\right)^2 \omega_0$$

Note that $R = \frac{R_0}{3}$, which can also be written as $3R = R_0$ or $3 = \frac{R_0}{R}$. Therefore, $\left(\frac{R_0}{R}\right)^2 = 9$.

$$\omega = \left(\frac{R_0}{R}\right)^2 \omega_0 = 9\omega_0$$

The final angular speed is 9 times greater than the initial angular speed. However, the problem **didn't** ask for angular speed. The problem asked for **period**. Use the equations $\omega = \frac{2\pi}{T}$ and $\omega_0 = \frac{2\pi}{T_0}$ (Chapter 16).

$$\omega = 9\omega_0$$
$$\frac{2\pi}{T} = 9\frac{2\pi}{T_0}$$

Divide both sides of the equation by 2π. The 2π's cancel out.

$$\frac{1}{T} = \frac{9}{T_0}$$

Cross multiply.

$$T_0 = 9T$$

Divide both sides of the equation by 9.

$$T = \frac{T_0}{9}$$

You should know that it normally takes $T_0 = 24$ hr for the earth to complete one revolution about its axis. (There is no need to convert.)

$$T = \frac{24}{9} = \frac{8}{3} \text{ hr} = 2.67 \text{ hr}$$

The final period is $T = \frac{8}{3}$ hr $= 2.67$ hr.

Example 205. As illustrated below, a 200-g pointlike object on a frictionless table is connected to a light, inextensible cord that passes through a hole in the table. A monkey underneath the table is pulling on the cord to create tension. The monkey is initially pulling the cord such that the pointlike object slides with an initial speed of 5.0 m/s in a circle with a 50-cm diameter centered about the hole. The monkey increases the tension until the pointlike object slides in a circle with a 20-cm diameter. Determine the final speed of the pointlike object.

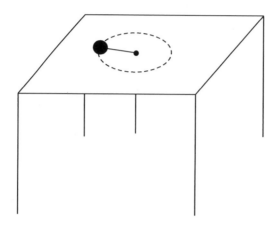

Solution. Since the net torque acting on the system (the pointlike object) equals zero (since the tension is directed toward the center of the circle, such that $\theta = 180°$ and $\sin 180° = 0$ in the equation for torque, $\tau = rF \sin \theta$), we may apply the law of conservation of angular momentum. The pointlike object is the only object in the system.

$$I_0 \vec{\omega}_0 = I \vec{\omega}$$

Apply the formula for the moment of inertia of a pointlike object (Chapter 32):

$$I_0 = mR_0^2 \quad , \quad I = mR^2$$

Substitute these expressions for moment of inertia into the equation for conservation of angular momentum.

$$mR_0^2 \omega_0 = mR^2 \omega$$

Use the formulas $v_0 = R_0 \omega_0$ and $v = R\omega$ (Chapter 16). Note that $R_0^2 = R_0 R_0$ and $R^2 = RR$.

$$mR_0 R_0 \omega_0 = mRR\omega$$
$$m(R_0\omega_0)R_0 = m(R\omega)R$$
$$mv_0 R_0 = mvR$$

Divide both sides of the equation by mass. The mass cancels.

$$v_0 R_0 = vR$$

Divide both sides of the equation by R.

$$v = \frac{R_0}{R} v_0 = \frac{50}{20}(5) = \frac{25}{2} \text{ m/s} = 12.5 \text{ m/s}$$

The final speed is $v = \frac{25}{2}$ m/s $= 12.5$ m/s.

WAS THIS BOOK HELPFUL?

A great deal of effort and thought was put into this book, such as:

- Breaking down the solutions to help make physics easier to understand.
- Careful selection of problems for their instructional value.
- Multiple stages of proofreading, editing, and formatting.
- Two physics instructors worked out the solution to every problem to help check all of the final answers.
- Dozens of actual physics students provided valuable feedback.

If you appreciate the effort that went into making this book possible, there is a simple way that you could show it:

Please take a moment to post an honest review.

For example, you can review this book at Amazon.com or BN.com (for Barnes & Noble).

Even a short review can be helpful and will be much appreciated. If you're not sure what to write, following are a few ideas, though it's best to describe what's important to you.

- Were you able to understand the explanations?
- Did you appreciate the list of symbols and units?
- Was it easy to find the information you were looking for?
- How much did you learn from reading through the examples?
- Would you recommend this book to others? If so, why?

Are you an international student?

If so, please leave a review at Amazon.co.uk (United Kingdom), Amazon.ca (Canada), Amazon.in (India), Amazon.com.au (Australia), or the Amazon website for your country.

The physics curriculum in the United States is somewhat different from the physics curriculum in other countries. International students who are considering this book may like to know how well this book may fit their needs.

GET A DIFFERENT ANSWER?

If you get a different answer and can't find your mistake even after consulting the hints and explanations, what should you do?

Please contact the author, Dr. McMullen.

How? Visit one of the author's blogs (see below). Either use the Contact Me option, or click on one of the author's articles and post a comment on the article.

www.monkeyphysicsblog.wordpress.com
www.improveyourmathfluency.com
www.chrismcmullen.wordpress.com

Why?
- If there happens to be a mistake (although much effort was put into perfecting the answer key), the correction will benefit other students like yourself in the future.
- If it turns out not to be a mistake, **you may learn something** from Dr. McMullen's reply to your message.

99.99% of students who walk into Dr. McMullen's office believing that they found a mistake with an answer discover one of two things:
- They made a mistake that they didn't realize they were making and learned from it.
- They discovered that their answer was actually the same. This is actually fairly common. For example, the answer key might say $t = \frac{\sqrt{3}}{3}$ s. A student solves the problem and gets $t = \frac{1}{\sqrt{3}}$ s. These are actually the same: Try it on your calculator and you will see that both equal about 0.57735. Here's why: $\frac{1}{\sqrt{3}} = \frac{1}{\sqrt{3}}\frac{\sqrt{3}}{\sqrt{3}} = \frac{\sqrt{3}}{3}$.

Two experienced physics teachers solved every problem in this book to check the answers, and dozens of students used this book and provided feedback before it was published. Every effort was made to ensure that the final answer given to every problem is correct.

But all humans, even those who are experts in their fields and who routinely aced exams back when they were students, make an occasional mistake. So if you believe you found a mistake, you should report it just in case. Dr. McMullen will appreciate your time.

VOLUMES 2 AND 3

If you want to learn more physics, volumes 2 and 3 cover additional topics.

Volume 2: Electricity & Magnetism
- Coulomb's law
- Electric field and potential
- Electrostatic equilibrium
- Gauss's law
- Circuits
- Kirchhoff's rules
- Magnetic field
- The law of Biot-Savart
- Ampère's Law
- Right-hand rules
- Magnetic flux
- Faraday's law and Lenz's law
- and more

Volume 3: Waves, Fluids, Sound, Heat, and Light
- Sine waves
- Oscillating spring or pendulum
- Sound waves
- The Doppler effect
- Standing waves
- The decibel system
- Archimedes' principle
- Heat and temperature
- Thermal expansion
- Ideal gases
- Reflection and refraction
- Thin lenses
- Spherical mirrors
- Diffraction and interference
- and more

ABOUT THE AUTHOR

Chris McMullen is a physics instructor at Northwestern State University of Louisiana and also an author of academic books. Whether in the classroom or as a writer, Dr. McMullen loves sharing knowledge and the art of motivating and engaging students.

He earned his Ph.D. in phenomenological high-energy physics (particle physics) from Oklahoma State University in 2002. Originally from California, Dr. McMullen earned his Master's degree from California State University, Northridge, where his thesis was in the field of electron spin resonance.

As a physics teacher, Dr. McMullen observed that many students lack fluency in fundamental math skills. In an effort to help students of all ages and levels master basic math skills, he published a series of math workbooks on arithmetic, fractions, algebra, and trigonometry called the Improve Your Math Fluency Series. Dr. McMullen has also published a variety of science books, including introductions to basic astronomy and chemistry concepts in addition to physics textbooks.

Dr. McMullen is very passionate about teaching. Many students and observers have been impressed with the transformation that occurs when he walks into the classroom, and the interactive engaged discussions that he leads during class time. Dr. McMullen is well-known for drawing monkeys and using them in his physics examples and problems, applying his creativity to inspire students. A stressed-out student is likely to be told to throw some bananas at monkeys, smile, and think happy physics thoughts.

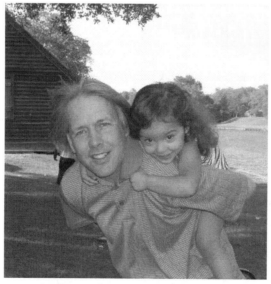

Author, Chris McMullen, Ph.D.

PHYSICS

The learning continues at Dr. McMullen's physics blog:

www.monkeyphysicsblog.wordpress.com

More physics books written by Chris McMullen, Ph.D.:
- An Introduction to Basic Astronomy Concepts (with Space Photos)
- The Observational Astronomy Skywatcher Notebook
- An Advanced Introduction to Calculus-based Physics
- Essential Calculus-based Physics Study Guide Workbook
- Essential Trig-based Physics Study Guide Workbook
- 100 Instructive Calculus-based Physics Examples
- 100 Instructive Trig-based Physics Examples
- Creative Physics Problems
- A Guide to Thermal Physics
- A Research Oriented Laboratory Manual for First-year Physics

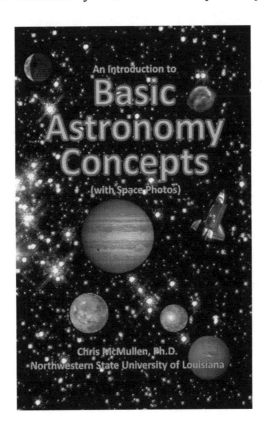

SCIENCE

Dr. McMullen has published a variety of **science** books, including:

- Basic astronomy concepts
- Basic chemistry concepts
- Balancing chemical reactions
- Creative physics problems
- Calculus-based physics textbook
- Calculus-based physics workbooks
- Trig-based physics workbooks

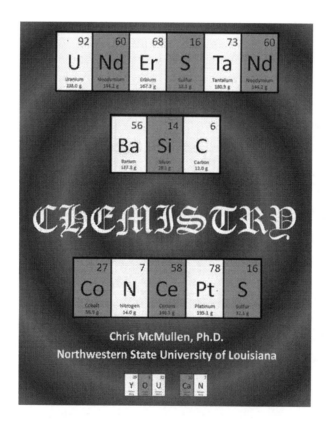

MATH

This series of math workbooks is geared toward practicing essential math skills:

- Algebra and trigonometry
- Fractions, decimals, and percents
- Long division
- Multiplication and division
- Addition and subtraction

www.improveyourmathfluency.com

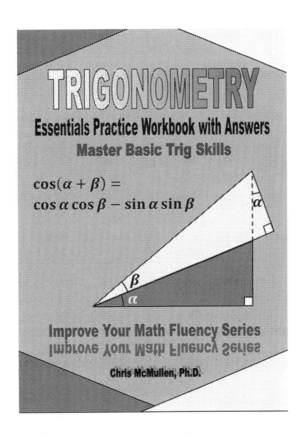

PUZZLES

The author of this book, Chris McMullen, enjoys solving puzzles. His favorite puzzle is Kakuro (kind of like a cross between crossword puzzles and Sudoku). He once taught a three-week summer course on puzzles. If you enjoy mathematical pattern puzzles, you might appreciate:

300+ Mathematical Pattern Puzzles

Number Pattern Recognition & Reasoning
- pattern recognition
- visual discrimination
- analytical skills
- logic and reasoning
- analogies
- mathematics

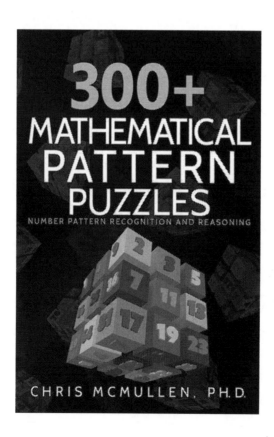

VErBAl ReAcTiONS

Chris McMullen has coauthored several word scramble books. This includes a cool idea called **VErBAl ReAcTiONS**. A VErBAl ReAcTiON expresses word scrambles so that they look like chemical reactions. Here is an example:

$$2\,C + U + 2\,S + Es \rightarrow S\,U\,C\,C\,Es\,S$$

The left side of the reaction indicates that the answer has 2 C's, 1 U, 2 S's, and 1 Es. Rearrange CCUSSEs to form SUCCEsS.

Each answer to a **VErBAl ReAcTiON** is not merely a word, it's a chemical word. A chemical word is made up not of letters, but of elements of the periodic table. In this case, SUCCEsS is made up of sulfur (S), uranium (U), carbon (C), and Einsteinium (Es).

Another example of a chemical word is GeNiUS. It's made up of germanium (Ge), nickel (Ni), uranium (U), and sulfur (S).

If you enjoy anagrams and like science or math, these puzzles are tailor-made for you.

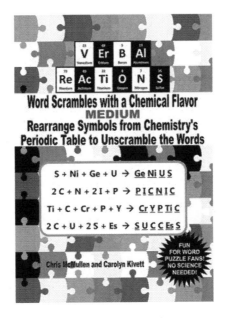

BALANCING CHEMICAL REACTIONS

$$2\,C_2H_6 + 7\,O_2 \rightarrow 4\,CO_2 + 6\,H_2O$$

Balancing chemical reactions isn't just chemistry practice.

These are also **<u>fun puzzles</u>** for math and science lovers.

Balancing Chemical Equations Worksheets
Over 200 Reactions to Balance
Chemistry Essentials Practice Workbook with Answers
Chris McMullen, Ph.D.

Made in the USA
San Bernardino, CA
29 October 2018